Universitext

Universitext

Editors (North America): J.H. Ewing, F.W. Gehring, and P.R. Halmos

Aksoy/Khamsi: Nonstandard Methods in Fixed Point Theory
Aupetit: A Primer on Spectral Theory
Bachem/Kern: Linear Programming Duality
Beneditti/Petronio: Lectures on Hyperbolic Geometry
Berger: Geometry I, II (two volumes)
Bliedtner/Hansen: Potential Theory
Booss/Bleecker: Topology and Analysis
Carleson/Gamelin: Complex Dynamics
Cecil: Lie Sphere Geometry: With Applications to Submanifolds
Chae: Lebesgue Integration
Chandrasekharan: Classical Fourier Transforms
Charlap: Bieberbach Groups and Flat Manifolds
Chern: Complex Manifolds Without Potential Theory
Cohn: A Classical Invitation to Algebraic Numbers and Class Fields
Curtis: Abstract Linear Algebra
Curtis: Matrix Groups
van Dalen: Logic and Structure
Das: The Special Theory of Relativity: A Mathematical Exposition
Devlin: Fundamentals of Contemporary Set Theory
DiBenedetto: Degenerate Parabolic Equations
Dimca: Singularities and Topology of Hypersurfaces
Edwards: A Formal Background to Mathematics I a/b
Edwards: A Formal Background to Mathematics II a/b
Emery: Stochastic Calculus
Foulds: Graph Theory Applications
Frauenthal: Mathematical Modeling in Epidemiology
Fukhs/Rokhlin: Beginner's Course in Topology
Gallot/Hulin/Lafontaine: Riemannian Geometry
Gardiner: A First Course in Group Theory
Gårding/Tambour: Algebra for Computer Science
Godbillon: Dynamical Systems on Surfaces
Goldblatt: Orthogonality and Spacetime Geometry
Gouvea: p-adic Numbers
Hahn: Quadradic Algebras, Clifford Algebras, and Arithmetic of Forms
Hlawka/Schoissengeier/Taschner: Geometric and Analytic Number Theory
Holmgren: A First Course in Discrete Dynamical Systems
Howe/Tan: Non-Abelian Harmonic Analysis: Applications of $SL(2,R)$
Humi/Miller: Second Course in Ordinary Differential Equations
Hurwitz/Kritikos: Lectures on Number Theory
Iversen: Cohomology of Sheaves
Jennings: Modern Geometry with Applications
Jones/Morris/Pearson: Abstract Algebra and Famous Impossibilities
Kelly/Matthews: The Non-Euclidean Hyperbolic Plane
Kempf: Complex Abelian Varieties and Theta Functions
Kloedern/Platen/Schurz: Numerical Solution of SDE Through Computer Experiments
Kostrikin: Introduction to Algebra
Krasnoselskii/Pekrovskii: Systems with Hysteresis
Luecking/Rubel: Complex Analysis: A Functional Analysis Approach

(continued after index)

Arlan Ramsay Robert D. Richtmyer

Introduction to
Hyperbolic Geometry

With 59 Figures

Springer-Verlag
New York Berlin Heidelberg London Paris
Tokyo Hong Kong Barcelona Budapest

Arlan Ramsay
Department of Mathematics
University of Colorado
Boulder, CO 80303
USA

Robert D. Richtmyer
Department of Mathematics
University of Colorado
Boulder, CO 80303
USA

Mathematics Subject Classifications (1991): 51-01, 51M09

Library of Congress Cataloging-in-Publication Data
Ramsay, Arlan.
    Introduction to hyperbolic geometry/Arlan Ramsay, Robert D.
Richtmyer.
        p.   cm. — (Universitext)
    Includes bibliographical references and index.
    ISBN 0-387-94339-0
    1. Geometry, Hyperbolic.   I. Richtmyer, Robert D.   II. Title.
QA685.R18   1994
516.9 — dc20   94-25789

Printed on acid-free paper.

Production managed by Natalie Johnson; manufacturing supervised by Gail Simon.
Photocomposed using the authors' TeX files.
Printed and bound by R.R. Donnelley and Sons, Harrisonburg, VA.
Printed in the United States of America.

9 8 7 6 5 4 3 2 1

ISBN 0-387-94339-0 Springer-Verlag New York Berlin Heidelberg
ISBN 3-540-94339-0 Springer-Verlag Berlin Heidelberg New York

Preface

This book is an introduction to hyperbolic and differential geometry that provides material in the early chapters that can serve as a textbook for a standard upper division course on hyperbolic geometry. For that material, the students need to be familiar with calculus and linear algebra and willing to accept one advanced theorem from analysis without proof. The book goes well beyond the standard course in later chapters, and there is enough material for an honors course, or for supplementary reading. Indeed, parts of the book have been used for both kinds of courses.

Even some of what is in the early chapters would surely not be necessary for a standard course. For example, detailed proofs are given of the Jordan Curve Theorem for Polygons and of the decomposability of polygons into triangles, These proofs are included for the sake of completeness, but the results themselves are so believable that most students should skip the proofs on a first reading.

The axioms used are modern in character and more "user friendly" than the traditional ones. The familiar real number system is used as an ingredient rather than appearing as a result of the axioms. However, it should not be thought that the geometric treatment is in terms of models: this is an axiomatic approach that is just more convenient than the traditional ones.

The book is appropriate as part of a special curriculum in undergraduate mathematics for exceptional students, designed to prepare them in three years for graduate-level mathematics. Our experience indicates that such students are able to learn mathematics more rapidly and in more depth than average undergraduate mathematics majors.

A principle that has guided the development is that of *mathematics in parallel*. Students can better see the interconnections if they hear about different subjects more-or-less simultaneously. For example, differential equations ought to be introduced right along with calculus; that makes calculus more interesting and more relevant. As another example, the beginnings of group theory should be included with linear algebra, because the permutation and rotation groups are already present, in effect, and it takes only a little further explanation (which the good students like) to lay the foundations of group theory. As a third example, some complex analysis is useful in calculus (for example power series in a complex variable); even Newton and Leibnitz used complex quantities (with trepidation).

The import of that idea for the present book is in the following:

(1) We have felt free to use methods of analysis, especially calculus and differential equations. In particular, we have used methods from differential geometry and have supplied two chapters on the relevant parts of that subject.

(2) We have assumed a knowledge of complex numbers and the complex plane, also of matrices and groups, up through homomorphism and normal subgroups.

(3) We have included a brief discussion of the connection with the Lorentz transformations of special relativity.

(4) We have taken the attitude that although students ought to know about the synthetic methods of geometric proof (they ought to have learned about it in high school geometry), and although they ought to know that there were gaps in Euclid which were bridged in the late 19th century, especially by Hilbert, and although they ought to know what a proof is and how to construct one, in fact the bright students don't need a whole semester of synthetic proving to achieve that. We have therefore adopted a set of axioms considerably less primitive than the axioms of Euclid or Hilbert. We feel in particular that in modern mathematics it is not necessary to derive the properties of the real number system from the axioms of geometry. (We have discussed that system independently in an appendix.) The purpose of our axioms is to tell the students what hyperbolic geometry *is* in terms that are familiar to them. However, the concepts that the axioms deal with should be intuitively geometrical. Geometry is not just an exercise in abstract logic, but the concepts deal with things that one can visualize to some extent.

(5) We have attempted to be more forthright about the role of models of the hyperbolic plane than some books are. Models were introduced originally for the purely logical purpose of proving that the axioms are self-consistent (assuming that the Euclidean axioms are self-consistent). One defines "points", "lines", "lengths" and "angles" in terms of certain (often rather strange looking) things in Euclidean geometry, and proves that the hyperbolic axioms are formally satisfied by the things in quotation marks. However, until one has proved that the axiom system is categorical, one cannot exclude the possibility that a given model may have additional properties that do *not* follow from the axioms (they merely don't contradict the axioms), while another model may have properties that contradict those of the first model, just as different groups have different properties even though every group satisfies all the axioms of group theory.

It is therefore not permitted to derive general theorems or formulas from a particular model rather than from the axioms (and the good students wouldn't let us get away with it, because they insist on understanding the logical connections in the subject). Our procedure is to derive the formulas of the hyperbolic plane from the axioms, without use of a model. To do that, we have first to establish the locally Euclidean nature of the hyperbolic plane: certain Euclidean laws are approximately satisfied by small figures, with a relative error that tends to zero as the size of the figures tends to zero. It seems to us that some books fail to make clear that the property of hyperbolic geometry follows from the hyperbolic axioms, not just from an accidental property of some model.

(6) After the formulas have been derived, we discuss models, and we prove that the axioms are categorical, i.e., that all models are isomorphic; the proof uses the formulas, so that categoricalness could not have been established (at least by this method) until the formulas are available. After categoricalness has been established, we can then derive further general results by use of any model.

(7) Since some of the very bright students have heard of Gödel's incompleteness theorem, we feel obliged to make some very brief remarks on the *apparent* conflict between that theorem and categoricalness, which says that the axioms of hyperbolic geometry are in a sense complete, for the purpose of describing the geometry.

(8) We have corrected a prevalent misunderstanding about astronomical parallax.

(9) We have tried to give an overview, not an encyclopedic sort of treatise. For example, most of the book deals with the two-dimensional cases, i.e., with the hyperbolic plane and the differential geometry of surfaces. Our discussion of the cases of more dimensions in Chapter 9 is admittedly quite sketchy, for we believe that most of the important concepts are found in the two-dimensional cases; once these are understood, the generalization contains no new difficulties.

The authors are grateful to our colleague Bill Reinhardt for much patience in discussing the relationship of Gödel's Incompleteness Theorem to the fact that the hyperbolic plane is unique up to isomorphism. Likewise, we thank our colleague Walter Taylor and his class in the fall term of 1993 for many suggestions for improving the ideas or the exposition. Elizabeth Stimmel was stellar in her performance turning the original typescript into TEX and her patience through numerous revisions. It is impossible to thank her enough. Finally, we thank Bruce Ramsay for invaluable assistance; he developed the software for drawing the figures and used it to produce them. He also gave crucial help in getting the figures into appropriate positions in the text in spite of the fiendish side effects of certain necessary TEX macros. He contributed much more than anyone can expect from a friend or a son.

Robert Richtmyer
Arlan Ramsay
Boulder, Colorado
August, 1994

Contents

Introduction

Over the centuries many attempts were made to prove the Parallel Postulate of Euclid. The discovery that an alternative is possible was a major breakthrough that lead to the development of hyperbolic geometry as a subject in its own right. The fact that the hyperbolic axioms are as consistent as the Euclidean axioms can be established by models within Euclidean geometry, but the fact that the hyperbolic axioms are categorical must be proved by careful development from the axioms. It is hoped that this book will encourage the development of geometric intuition and also the recognition that mathematics itself is not compartmentalized in spite of the fact that names are given to various branches. The close relationship between analysis and geometry allows the use of analysis to simplify many derivations and proofs.

A Bit of History

Investigations of geometry during the nineteenth century achieved two things: (1) It was discovered that although Euclidean geometry represented one of the great achievements of all time by introducing logical methods, it was not fully rigorous; the axioms were often rather informal, and some additional things were assumed without proof. Euclidean geometry was therefore reworked, especially by David Hilbert, in the last few years of the century. (2) It was found that if the Euclidean parallel axiom was replaced by a different axiom, which allows more than one parallel to a given line through a given point, a new kind of geometry resulted, now called hyperbolic geometry, and it was proved, by means of models, that the new geometry was no less consistent than Euclidean geometry. (Spherical geometry goes in the other direction, in that there are no parallels at all, but then other axioms have to be changed; in a sense, hyperbolic geometry is closer to Euclidean geometry than are other non-Euclidean geometries.)

From ancient until modern times, geometry preceded analysis; the rigorous and logical basis of mathematics was in geometry. Today, analysis precedes geometry; the rigorous basis is in the set-theoretical foundations of real analysis. In ancient mathematics, the properties of the real number system were derived from the axioms of geometry. Today, they are derived from set theory, as in Chapter 5 of the book by Herbert B. Enderton (*Elements of Set Theory,* Academic Press, 1977).

What Are Axioms For?

The axioms of a subject have traditionally served two purposes. The first is to provide a rigorous basis for the subject, in which the axioms assert just as little as possible without proof, and everything else results by deduction according to rules of logic. The second purpose is to define or describe the structures or objects to be discussed; the axioms of group theory single out a class of objects called groups, and so on.

We now have easier ways of being rigorous in geometry than were available in the times of the ancient Greeks. Hence, the first of the above purposes has lost its value, except in a historical context. Today, analysis has an independent existence and can be used in describing geometry. We can now use the properties of the real number system and the laws of arithmetic and algebra and calculus in addition to the rules of logic. Doing that has two advantages: first, it makes things rigorous, because the foundations of analysis are based on modern set theory and logic, which are more rigorous than the geometry of Euclid; second, the methods of analysis make the derivations and calculations easier.

Consequently, the approach called synthetic geometry has lost most of its importance, from the point of view of understanding the structure of mathematics. On the other hand, some of the classical theorems can be regarded as almost as fundamental, for telling us what geometry is, as the axioms. For example, when Hilbert's axioms of incidence, betweenness, congruence, and continuity, together with a fair number of propositions derived from them, are applied to the ordered set of points that constitute a straight line, the upshot is that that set of points is homeomorphic, and in fact isometric, with the set of real numbers in the real number system $\mathbb{R}$. That is equivalent to quite a lot of statements, but those statements are logically consistent, because of the (assumed) logical consistency of the real number system $\mathbb{R}$. Anyone who understands the real number system knows what a "continuum" is, and the above statement says just that the set of points on a line is a continuum. There is no reason why we should not take that statement as an axiom. Certain conclusions are immediate; for example, Pasch's theorem and the crossbar theorem, when carefully formulated, can be simplified by using the intermediate-value theorem for continuous functions (see Section 2.3).

Another example concerns the triangle inequality, which says that if A, B, C are any points, and if $|AB|$ denotes the distance between A and B, and so on, then $|AC| \leq |AB| + |BC|$. This inequality (which can be proved as a theorem from Hilbert's axioms) embodies the intuitive geometrical notion that a straight line is the shortest path between two points, and it makes the plane (Euclidean or hyperbolic) into a metric space, a thing familiar to students of analysis. So we take the triangle inequality as one of the axioms of hyperbolic geometry (but of course we point out that it doesn't hold in spherical geometry).

The difference is rather subtle. The axioms of a subject are statements about otherwise undefined things (points, lines, etc.) and are used for building up the subject by logical deduction. But, to make a *statement* about those things, one has to have a language, and the words of the language have to be understood and defined in advance. The language we have chosen includes words and concepts from real analysis.

In summary, although axioms are still essential for specifying a particular kind of geometry, such as hyperbolic plane geometry, they don't need to be as primitive and mutually independent as earlier; the axiom system can be more "user-friendly"; its only purpose is to define the geometry in terms that are familiar to the student.

What Are Models For? (Consistency and Categoricalness)

In some books, the role of models of the hyperbolic plane is not made clear. The models serve primarily a *logical* purpose, in proving the consistency of the axioms (assuming that the Euclidean geometry is consistent); they don't "look like" the hyperbolic plane.

Two important notions about the axioms of a subject are consistency and categoricalness. The axioms are *consistent* if they can never lead to a contradiction. That is, of course, always essential. Second, a set of axioms is *categorical* if it contains enough axioms to specify the system completely and uniquely. Models play a role for both these ideas.

If one can find a collection of objects and relations among those objects such that all the axioms of a system are satisfied, one says that one has a *model* of the system. If the objects and relations that appear in the model are based on a simpler set of axioms that one believes in advance to be consistent, then the consistency of the given axiom set is thereby established. An example is provided by complex numbers. In the 18th century many mathematicians feared that to postulate the existence of a "number" i such that $i^2 = -1$ and so on was dangerous and could lead to contradictions. That fear was proved groundless, early in the 19th century, independently by Gauss and Hamilton, who showed how to construct a model of the complex number system, in which each complex number is represented by a pair (a, b) of real numbers, with $(a, b) + (c, d)$ defined as $(a + c, b + d)$ and $(a, b)(c, d)$ as $(ac - bd, ad + bc)$. The complex number system was thereby proved to be consistent, on the assumption that the real number system is consistent, and it became henceforth irrelevant whether one wrote (a, b) or $a + ib$.

The logic behind such a proof of consistency is this: The axioms of a system deal with undefined objects (e.g., "points" in geometry), which in themselves contain no contradictions, and in fact have no properties at all, except as furnished by the axioms. Secondary things are then built up from them by definitions and the axioms. If any contradiction results, it must involve things whose very existence is guaranteed by the axioms. Therefore, such a contradiction would have to appear also in any model, if the model satisfies all the axioms.

However, consistency does not imply categoricalness. Stated informally, the axioms of a system may not say all that *could have been said*. In that case, different models can have different properties. One model may, in effect, be following also a possible further axiom and a second model another, and the two further axioms may not be consistent with each other. An example of a noncategorical set of axioms is the axioms of a group. We have many "models" of the abstract notion of a group (many specific groups); each of them satisfies all the axioms of a group, but what is true for one is often false for another. In such a case, it is clearly invalid to "prove" theorems in the subject from a model. An axiom system is categorical if all models are abstractly identical, i.e., isomorphic. For a simple example, all groups of order 3 are isomorphic.

For another example, there are many models of the real number system; models using decimals, or using Cauchy sequences of rational numbers, or using Dedekind cuts. The appropriate axiom system is that of a complete ordered field, and a proof is given, at least in the better books on advanced calculus, that any two models of a complete ordered field are isomorphic; hence, any model can be used for deriving properties of the real number system. See the Appendix to Chapter 1.

The axioms of hyperbolic plane geometry are in fact categorical, but the proof of that is decidedly not elementary — see Section 7.7.

There is a subtle distinction between categoricalness and completeness of an axiom system, which has to be made clear, to avoid conflict with Gödel's incompleteness theorem. See Section A.6 in the Appendix to Chapter 1.

In the absence of a proof of categoricalness, it would be unjustified to derive results from a model, as is done in some books (or to "prove" categoricalness by means of results that have been derived from a model). That misunderstanding has a partly historical origin; the inventors of the models of hyperbolic geometry, mainly Beltrami, Klein, and Poincaré, were concerned, at the time, only in proving consistency. Several investigators spoke of a "sphere of imaginary radius"; they replaced the radius r by ir in the formulas for spherical geometry. That is a convenient method of making conjectures, but, although the resulting formulas satisfy the axioms, it doesn't prove that they follow uniquely from the axioms. The idea of deriving the geometry rigorously from a set of axioms came later, and was due largely to Hilbert.

We shall derive all results from the axioms. We shall not derive anything using a model until the axioms have been proved to be categorical.

What Is Geometry? (The Problem of Visualization)

The predominant role of analysis in modern mathematics does not mean that we should give up the axiomatic approach and treat geometry merely as a branch of analysis. The notions of geometry go, in a sense, beyond the notions of analysis, in that they are things that we "visualize." Although we must keep in mind the limitations of diagrams, and so on, the ability to visualize is a human ability that should be encouraged rather than suppressed, in the teaching of mathematics. Our impression from teaching talented young students is that they *can* visualize the hyperbolic plane, in a sense. From that point of view the main models, those of Beltrami, Klein, and Poincaré, are unsatisfactory for intuitive geometrical visualization. The homogeneity and isotropy of the hyperbolic plane are not at all evident in the models; one can of course define the isometries as certain nonlinear transformations of the models, but under them, congruent figures don't even look alike, and besides, the hyperbolic plane extends to infinity in all directions and the models don't. From that point of view, the models should be regarded as serving a *logical* not *geometrical* purpose, namely, to establish the consistency of the axioms.

In the first six chapters, we describe the hyperbolic plane in *geometrical* terms independent of the models. The models are introduced only in Chapter 7. The proof of consistency of the axioms and the proof of categoricalness are given in Chapter 7. In consequence of the latter proof, it is permissible thereafter to derive general results from any model, and certain results are indeed thus derived in subsequent chapters. The proof of consistency could have been given earlier, for example by introduction of one of the models earlier, but the proof of categoricalness appears to require some of the results of the first six chapters.

The Role of Analysis in Geometry (Differential Geometry)

In the approach we follow, one starts with a set of axioms roughly equivalent to Hilbert's axioms (but made more "user-friendly" as described above), and then one gets as rapidly as possible, by synthetic methods, to the point where the methods of analysis can be used (analytic geometry, calculus, differential equations, and differential geometry). For example, one can show that the function for the angle of parallelism, $\Pi(y)$, satisfies a differential equation, which is easy to solve. (See Exercise 4 in Section 6.3.) In contrast to derive the same formula by purely synthetic methods requires going into three-dimensional hyperbolic geometry, using horospheres and making many complicated constructions in a very long process.

The key to our procedure is the locally Euclidean nature of the hyperbolic plane for very small figures. For example, if a right triangle in the Euclidean plane has hypotenuse c and angle α opposite side a, then $a = c\sin\alpha$; in the hyperbolic plane, we shall find that $a = (c\sin\alpha)(1 + \varepsilon)$, where ε is a quantity that goes to zero at a certain rate as the dimensions of the triangle go to zero. By taking limits as certain quantities go to zero, we arrive at derivatives and other notions of analysis. This is in analogy with calculus itself, which used to be called "infinitesimal calculus"; to understand it, one has to understand the "infinitesimals."

The chief difficulty concerns a sort of smoothness of the geometry. The topology is easy (i.e., the topology associated with the metric). As stated above, each line is homeomorphic to $\mathbb{R}$, and the plane is homeomorphic to $\mathbb{R} \times \mathbb{R}$, but more than that is difficult. For example, one of the axioms shows that if a and b are the lengths of two sides of a triangle, and γ is the included angle, then the third side c is given by a function $\phi(a, b, \gamma)$; the axiom says nothing about the nature of that function. It is easy to prove that that function is continuous in each of its variables, and that it satisfies a Lipschitz condition in each variable, but to prove that it is even once differentiable (it is in fact analytic) is more difficult. This is similar to the difficulty of Hilbert's fifth problem, which says that if a continuous manifold contains a group structure that is continuous on the manifold, then it follows, without any more axioms, that the manifold is differentiable and even analytic. That was conjectured by Hilbert in 1900 and proved only in the 1950's. The group axioms somehow have the power to enforce analyticity, even though those axioms are quite simple and "innocent" in appearance. Similarly, the axioms of geometry seem to imply a smoothness of space. The connection between these two things is probably even rather close, because the axioms of the geometry imply the existence of an important group, namely, the group of isometries described in Section 3.6.

As soon as smoothness has been established, a metric space can be completely determined by giving its so-called metric tensor. Then, analytic methods can be used. The "lines" are the geodesics, to be determined as in Section 5.4; distances and angles are also determined in terms of the metric tensor.

The smoothness or locally Euclidean nature of the hyperbolic plane is established in Chapter 4. There are various ways in which that can be done. We have chosen a method due to Lobachevski, which has the advantage that it gives us an opportunity to introduce hyperbolic 3-space and a remarkable

7969

778

7777

property of the horosphere, an important kind of surface in that space. The smoothness enables us to derive explicit formulas for triangles and so on in the hyperbolic plane; that is done in Chapter 6.

Outline of the Following Chapters

The first chapter contains a system of axioms and definitions designed to be a specification of hyperbolic plane geometry in terms of things familiar to the student, including things in analysis.

Chapter 2 reviews relevant things from Euclid (so-called neutral theorems), and adds some neutral information suggested by more recent mathematics.

Chapter 3 gives the qualitative picture of the hyperbolic plane that results from those axioms. Surprisingly many general things can be discussed without use of quantitative formulas, including things having to do with isometries, tilings, and invariant curves.

Chapter 4 prepares the way for the derivation of formulas. The key is the locally Euclidean nature of the hyperbolic plane for very small figures, which makes it possible to discuss differentiation and other limiting processes.

Chapter 5 contains those parts of differential geometry that are relevant to the hyperbolic plane.

Chapter 6 gives actual formulas for many things: for right triangles, for the angle of parallelism, for circular arc length, for straight lines in polar coordinates, and so on. According to the procedure followed here, these formulas are derived from the axioms, that is, from the geometrical specification, rather than from a model, because no proof of categoricalness has yet been given; in fact, that proof, given later, uses some of the formulas derived here.

The next two chapters discuss various models of the hyperbolic plane and the matrix representation of the isometry group. The isometry group plays a prominent role. According to the ideas of Klein's Erlanger Program, that group really characterizes the geometry, in a sense. It is shown that the Euclidean and hyperbolic isometry groups have different structures. In the former, the set of all translations is a subgroup, in fact, a two-dimensional normal subgroup. In the hyperbolic case, there is no such subgroup, and the translations don't constitute a subgroup at all.

A consequence is that, although in the Euclidean case any number of pure translations, performed one after another always give another pure translation, rotations can result in the hyperbolic case. One can find two pure translations whose resultant is a pure rotation. That same effect appears in special relativity, where the resultant of several pure Lorentz transformations or "boosts" can result in a rotation; that is of importance in atomic theory. It is shown that the hyperbolic isometry group is identical with the group of Lorentz transformations in two space variables and time.

Chapter 9 discusses hyperbolic geometry in more dimensions.

Chapter 10 discusses the connection with the Lorentz group of special relativity and relativistic velocity space. That requires a section on Lorentz transformations, which, we find, good students can understand even though they may not have studied relativity theory formally.

Chapter 11 discusses the problem of constructions with straightedge and compass in the hyperbolic plane. The purpose is of course not to consider practical draftsmanship, but to show what light the construction problem throws on the nature of the geometry. At the end of that chapter we come, in fact, to the idea that the constructible figures determine a kind of geometry based on all the axioms of the hyperbolic plane, except that the completeness axiom is omitted. In that geometry there is no analysis, because the notion of convergence of a sequence of points cannot be used, but all the other aspects of the geometry are there.

For many applications, especially to other branches of mathematics, perhaps the most important single result of the entire subject is that the hyperbolic plane, with all the features discussed in Chapter 3, is completely represented by the Poincaré half-plane model, which consists of the upper half of the x, y plane $(y > 0)$ with a Riemannian metric expressed by the differential line element $ds^2 = y^{-2}(dx^2 + dy^2)$ (from which the geometry follows). From that point of view, a main purpose of these chapters is to show how that complete representation comes about.

Chapter 1

Axioms for Plane Geometry

The axioms systems of Euclid and Hilbert were intended to provide everything needed for plane geometry without any prior development. The axioms of Hilbert include information about the lines in the plane that implies that each line can be identified with the structure commonly called the "real numbers" and denoted by $\mathbb{R}$. Euclid's axioms also include information of that kind but the meaning of the "real numbers" may not have been the same in that era as it is now. In both cases, geometry was taken as more fundamental than the real number system. Instead we are going to use $\mathbb{R}$ as an ingredient in laying the foundations of hyperbolic plane geometry. (See the Appendix to this chapter for a discussion of the real number system, its properties, its consistency and its uniqueness.) Our axiom system is equivalent to that of Hilbert for the hyperbolic plane, and following the laying of the foundations in this chapter we proceed rigorously with the development of its properties, its consistency and its uniqueness, in later chapters.

1.1 The Axioms, Definitions and Remarks

The n-dimensional Euclidean space is denoted by $\mathbb{E}^n$ and the n-dimensional hyperbolic space by $\mathbb{H}^n$. In this chapter, we deal with the planes $\mathbb{E}^2$ and $\mathbb{H}^2$; the symbol $\mathbb{P}$ will stand for either of them. Definitions are made by simply putting the term to be defined in italics. *Remarks* follow from the axioms with very little proof (which we give or at least outline); they could in principle have been put into the next chapter as neutral theorems, but we prefer to put them here because of the light that they throw on the axioms and definitions. Comments in square brackets [ ] are not a part of the formal development, but indicate things to be proved later or elsewhere.

The plane $\mathbb{P}$ is a set of previously undefined things called *points*, usually denoted by $A, B, C, \ldots$, and certain subsets of $\mathbb{P}$ called *lines* (or *straight lines*), usually denoted by $\ell, m, n, \ldots$, subject to the axioms given below. All the axioms until the last (Axiom 7) apply to both the Euclidean and the hyperbolic planes; there are two forms of Axiom 7, one for each of the two kinds of geometry, Euclidean and hyperbolic.

If ℓ is a line, that is, is one of the special subsets of $\mathbb{P}$ mentioned above, and A is a point in that subset, we say that A *lies on* ℓ or ℓ *passes through* A. These are just different ways of saying that A belongs to the subset ℓ.

Axiom 1: If A and B are distinct points (i.e., the symbols A and B denote *different* points), then there is one and only one line that passes through them both. This line is denoted by $\overline{AB}$.

Axiom 2: There is a function $|PQ|$ defined for all pairs of points P, Q, such that $|PQ| = 0$ if P and Q are the same point, and is a positive number otherwise; it is such that $|PQ| = |QP|$ and such that the triangle inequality

$$|AC| \leq |AB| + |BC|$$

holds for all triples of points A, B, C.

This axiom makes the plane into a so-called *metric space*; $|PQ|$ is called the *distance* between points P and Q.

Remark 1: By transposing first one then the other of the terms from the right to the left and suitably permuting the symbols, we get the more complete form

$$\big||AB| - |BC|\big| \leq |AC| \leq |AB| + |BC|, \tag{1.1-1}$$

where the long vertical lines on the left denote absolute value.

Comment: It will be proved in the next chapter that the right-hand inequality in this formula is equality if and only if A, B, C are collinear and B is in the segment AC; the left-hand inequality is equality if and only if A, B, C are collinear and B is *not* in the segment AC. See Corollary 2 to Theorem 2.2 in Section 2.3 of the next chapter.

Axiom 3: For each line ℓ, there is a one-to-one mapping, x, from the set ℓ to the set $\mathbb{R}$ of real numbers, such that if A and B are any points on ℓ, then $|AB| = |x(A) - x(B)|$.

Remark 2: The mapping given by $x(A)$ is not unique. If $x(A)$ is one such mapping, then any mapping of the form

$$y(A) = x(A) + \text{const} \qquad \text{or} \tag{1.1-2}$$
$$y(A) = -x(A) + \text{const} \tag{1.1-3}$$

is another one that has the same properties. In the language of topology, this definition makes each line into a *continuum* homeomorphic to the real number system $\mathbb{R}$. In the language of metric spaces, this definition makes each line *isometric* to $\mathbb{R}$.

Some Definitions: The mapping $A \to x(A)$ is a *coordinatization* of the line, and the number $x(A)$ is the *coordinate* of the point A. If A and B are given points on ℓ, the subset of ℓ consisting of all points P such that $x(A) \leq x(P) \leq x(B)$ or $x(B) \leq x(P) \leq x(A)$ [depending on whether $x(A) < x(B)$ or $x(B) < x(A)$] is a *segment* of ℓ, denoted by AB, and the number $|AB|$ is its *length*; A and B are its *endpoints*. When a coordinatization is given we say that ℓ is a *directed line* (we can put an arrowhead on a line in a figure to indicate the direction of increasing coordinate x). A re-coordinatization of type (1.1-2) leaves the direction unaltered, while (1.1-3) reverses it. If Z is a point on ℓ, the set of points A such that $x(A) > x(Z)$ is a *ray* $\overrightarrow{k}$ with Z as its *origin* or *initial point*. (Note that Z is not itself a point of the ray.) The set of points A such that $x(A) < x(Z)$ is called the *opposite ray* and is often denoted by $\overleftarrow{k}$. [Just which ray is denoted by $\overrightarrow{k}$ and which by $\overleftarrow{k}$ is of course arbitrary, because the re-coordinatization (1.1-3) interchanges them.] A ray from a point Z will often be denoted by $\overrightarrow{ZP}$, where P is a point on the ray, it being understood, of course, that the ray contains also the points "beyond" P.

The next axiom is connected with the two-dimensional or planar nature of $\mathbb{P}$ by asserting that each line separates it into two half-planes.

Axiom 4: If ℓ is any line, there are two corresponding subsets HP_1 and HP_2, called *half-planes*, such that the sets ℓ, HP_1 , and HP_2 are disjoint, and their union is all of $\mathbb{P}$, such that
(a) if P and Q are in the same half-plane, the segment PQ contains no point of ℓ , while
(b) if P and Q are in opposite half-planes, the segment PQ contains a (single) point of ℓ . (See Fig. 1.1a.)

We say that P and Q are on the *same side* of or on *opposite sides* of ℓ or of a ray (meaning of the line that contains that ray).

Rays $\overrightarrow{h}$ and $\overrightarrow{k}$ with a common origin Z constitute an *angle*, denoted by $\angle \overrightarrow{h}, \overrightarrow{k}$ or $\angle \overrightarrow{k}, \overrightarrow{h}$; Z is its *vertex*. If, furthermore, B is a point on $\overrightarrow{h}$ and C a point on $\overrightarrow{k}$, the angle is also denoted by $\angle BZC$ or $\angle CZB$ (or sometimes simply by $\angle Z$, if only one angle has been mentioned in a given discussion with its vertex at Z). If $\overrightarrow{h}$ and $\overrightarrow{k}$ are neither the same ray nor opposite rays, the *interior* of the angle is the set of all points on the same side of $\overrightarrow{h}$ as C and on the same side of $\overrightarrow{k}$ as B; it is shown shaded in Fig. 1.1b. The *interior* of a triangle ABC consists of the points that are on the same side of $\overline{AB}$ as C, on the same side of $\overline{BC}$ as A, and on the same side of $\overline{AC}$ as B; hence it is the intersection of the interiors of the three angles of the triangle.

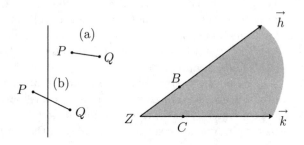

Fig. 1.1a.   **Fig. 1.1b.**

Remark 3: If ℓ amd m are lines that intersect at a point Z and x is a coordinate function on ℓ, then points P, Q of ℓ are in the same half-plane determined by m if and only if $x(P)$ and $x(Q)$ are on the same side of $x(Z)$. Thus half-planes are compatible with rays.

Remark 4: The above definition is independent of the choices of B on $\overrightarrow{h}$ and C on $\overrightarrow{k}$; the lines containing those rays are distinct (because the rays are neither the same nor opposite); hence, those lines have no intersection except at Z. Hence, if B and B' are any two points of $\overrightarrow{h}$ different from Z, then they are on the same side of $\overrightarrow{k}$, so if P is any point, P and B are on the same side of $\overrightarrow{k}$ if and only if P and B' are on the same side. A similar remark applies to points C and C' on $\overrightarrow{k}$.

Remark 5: If another ray $\overrightarrow{j}$ with origin Z has a point P in the interior of $\angle \overrightarrow{h}, \overrightarrow{k}$, then all points of $\overrightarrow{j}$ are in that interior, because points P and P' on $\overrightarrow{j}$ are on the same side of $\overrightarrow{k}$ and also on the same side of $\overrightarrow{h}$. We then say simply that the ray $\overrightarrow{j}$ *is in the interior of the angle.* According to the crossbar theorem, Corollary 1 of Theorem 2.3, points like B and C are then on *opposite* sides of j (this may seem obvious, but needs proof).

If two lines intersect at Z, if $\overrightarrow{h}$ and $\overleftarrow{h}$ are the opposite rays starting at Z that make up the first line, and $\overrightarrow{k}$ and $\overleftarrow{k}$ are opposite rays that make up the other line, then the angles $\angle \overrightarrow{h}, \overrightarrow{k}$ and $\angle \overrightarrow{h}, \overleftarrow{k}$ are called *supplementary angles*, and the angles $\angle \overrightarrow{h}, \overrightarrow{k}$ and $\angle \overleftarrow{h}, \overleftarrow{k}$ are called *vertical angles*. The four angles at Z thus appear in pairs (four pairs) of supplementary angles and in pairs (two pairs) of vertical angles. See Fig. 1.1c.

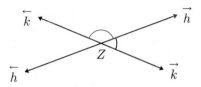

Fig. 1.1c.

In analogy with Axiom 3, which says that the set of all points on a line forms a continuum, the next axiom says that the set of all angles with a given vertex and in a given half-plane also forms a continuum.

Axiom 5: (Measures of angles) For each angle $\angle \overrightarrow{h}, \overrightarrow{k}$ there is a number $\angle \overrightarrow{h}, \overrightarrow{k}^{\mathrm{rad}}$ in the interval $[0, \pi]$, called the (radian) *measure* of the angle, such that

(a) if $\overrightarrow{h}$ and $\overrightarrow{k}$ are the same ray, the measure is 0; if they are opposite rays, the measure is π;

(b) the sum of the measures of an angle and its supplement is π;

(c) if $\overrightarrow{j}$ is in the interior of an angle $\angle \overrightarrow{h}, \overrightarrow{k}$, then $\angle \overrightarrow{h}, \overrightarrow{j}^{\mathrm{rad}} + \angle \overrightarrow{j}, \overrightarrow{k}^{\mathrm{rad}} = \angle \overrightarrow{h}, \overrightarrow{k}^{\mathrm{rad}}$;

(d) if a ray $\overrightarrow{k}$ from a point Z lies in a line ℓ , then in each half-plane bounded by ℓ , the set of all rays $\overrightarrow{j}$ from Z is in a one-to-one correspondence with the set of all real numbers α in $(0, \pi)$ in such a way that α is equal to the measure $\angle \overrightarrow{j}, \overrightarrow{k}$ ^{rad} of the angle $\angle \overrightarrow{j}, \overrightarrow{k}$; and

(e) under the conditions of part (d), if the ray $\overrightarrow{j}$ is determined as $\overrightarrow{ZP}$, where P is a point on $\overrightarrow{j}$, as in Fig. 1.1d, then the angle α depends continuously on P; that is, if P' is a variable point and α' the corresponding angle, then $\alpha - \alpha' \to 0$ as $|PP'| \to 0$.

[Note: Part (e) can be proved from the other parts and the other axioms; hence it might have been put into the next chapter as one of the neutral theorems, but we prefer to include it here as part of the axiom.]

In much of the following, a less formal notation will be used in which an angle is denoted by a single letter (usually a small Greek letter); for instance, α might denote either the symbol $\angle \overleftrightarrow{h}, \overrightarrow{k}$ or the number $\angle \overrightarrow{h}, \overrightarrow{k}^{\mathrm{rad}}$. Angles will sometimes be expressed in degrees rather than radians, and the symbol $\angle \overrightarrow{h}, \overrightarrow{k}^{\circ}$ will be used.

Comment: We assume the reader is familiar with radians, but it must be pointed out that the numerical value of π has no geometrical significance

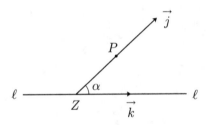

Fig. 1.1d.

in hyperbolic geometry until Chapter 6; in particular, it is not the ratio of the circumference of a circle to the diameter, except in the limit as the diameter tends to zero.

Now, all segments have lengths and all angles have measures. Two segments with the same length are called *congruent segments*, and two angles with the same measure are called *congruent angles*. A *right angle* is one that is congruent to its supplement.

Remark 6: Supplements of congruent angles are congruent (because they have the same measure, according to the axiom), and vertical angles are congruent (because they are both supplements of one of the angles in between). A right angle has radian measure $\pi/2$, according to part (b).

We now come to the axiom that completes the determination of the geometry, except for the final axiom that distinguishes between the Euclidean and hyperbolic cases. A *triangle* is a set of three noncollinear points A, B, C and is denoted by ABC or by $\triangle ABC$. The points A, B, C are its *vertices*, the segments AB, BC, and AC are its *sides*, and the angles at A, B, C are its *angles*. We shall often denote the angles at A, B, C by α, β, γ, respectively, and the lengths of the sides opposite those angles by a, b, c, respectively.

Axiom 6: (SAS or side-angle-side criterion for congruence). If two sides and the included angle of a first triangle are congruent respectively to two sides and the included angle of a second triangle, then the triangles are congruent. That is, the remaining sides are congruent and the remaining angles are congruent in pairs.

Let us say this a little more precisely. A triangle ABC is congruent to another triangle (say DEF) if there is a one-to-one mapping σ from the vertices of the first triangle to those of the second such that the angles at A and $\sigma(A)$ are the same, the length of side AB is the same as the length of the side $\sigma(A)\sigma(B)$, and the length of AC is the same as the length of $\sigma(A)\sigma(C)$. The axiom says that then the angles at B and $\sigma(B)$ are the same, the angles at C and $\sigma(C)$ are the same, and the lengths of the segments BC and $\sigma(B)\sigma(C)$ are the same. (In the next chapter, there is an application of this axiom when the two triangles are the same triangle, and the mapping σ simply interchanges two of its vertices.)

This axiom establishes the existence of a (real) function $\phi(a, b, \gamma)$ such that the equation $c = \phi(a, b, \gamma)$ gives the length c of the third side in terms of the lengths a, b of two sides and their included angle, for any triangle, regardless of its location and orientation in the plane. Since the axiom says nothing about the order of the sides (i.e., whether a, b, c goes clockwise or counterclockwise round the triangle), we see that $\phi(a, b, \gamma) = \phi(b, a, \gamma)$ and that any triangle is congruent to its mirror image. The other familiar criteria for congruence (ASA, SAA, SSS) are discussed in the next chapter, where they are derived from the axioms of the present chapter.

It follows from the foregoing axioms, via the so-called alternate interior angles theorem (see next chapter), that through any point there is always at least one line parallel to a given line, by which we mean one that does not intersect the given line. Whether there are more than one is the subject of the final axiom.

Axiom 7a: (Euclidean parallel axiom). Given any line and any point not on it, there is only one line through the given point that never intersects the given line.

Axiom 7b: (Hyperbolic parallel axiom). There exists a line and a point not on the line such that there are at least two lines through that point that do not intersect the given line.

It will be seen in Chapter 3 that, in the hyperbolic plane, there are infinitely many lines through that point that do not intersect that line, and that the same is true also for any other given line and any other given point not on it.

1.2 Comments on the Unit of Length

The usual concept of distance or length, as it applies in the physical world, requires that we choose a unit of length. To say that the distance between two points is 3.5 is meaningless until we specify the unit. Lengths of 3.5 centimeters, 3.5 inches, 3.5 miles, and 3.5 light-years are different. The choice of the unit is a matter of physics, not mathematics. We must have a meter stick of wood (or a precious metal) or (in the case of the light-year) a planet revolving about a sun (to determine the year). There is nothing in Euclidean geometry to single out one choice of the unit as being in any way more special than other choices.

In the hyperbolic plane, the situation is slightly different. In the first place, the choice of the unit is still arbitrary, in the sense that it is not specified in the axioms, but there are various ways of singling out one choice as being geometrically unique. The axioms assert the existence of a real function $|PQ|$ defined for pairs of points, and on each line a real function $x(P)$ called a coordinate, which is related to the function $|PQ|$ in a specified way — but the axioms don't say that those functions are unique. In fact, it follows from the axioms that there are many possible choices of those functions. If $|PQ|$ and $x(P)$ are one choice, and if a is any positive number, then the functions $|PQ|' = a|PQ|$ and $x'(P) = ax(P)$ are another choice that also satisfies the axioms. With each choice there exists a segment AB such that $|AB| = 1$, which may be called a unit segment, but that can be any segment, depending on the value of a.

In the second place, formulas will be derived in Chapter 6 from the axioms for various functions, for example, the function $\phi(a, b, \gamma)$ which gives the length of one side of a triangle when the other two sides and the included angle are known. All of those formulas contain a constant K, a positive number, whose value depends on the unit of length, and there is a unique choice of the unit such that $K = 1$. From Chapter 6 on, we shall often assume that choice of unit, to simplify the writing of the formulas.

Some authors say, in connection with the problem of categoricalness, that two models of the hyperbolic plane are isomorphic if and only if they have the same value of K. That amounts to assuming that the unit of length has been specified in the model, and then the length $1/K$ is regarded as determining a length scale so that some models have larger features than corresponding features of other models. For many applications of hyperbolic geometry, it is useful to consider different models with different length scales, but the length scale must be regarded as a property of the model, not of the abstract hyperbolic plane. At least when the axioms are the ones we have chosen, all models are isomorphic with respect to things uniquely determined by the axioms.

According to the methods of differential geometry, discussed briefly in Chapter 5, $1/K$ may be regarded as a "curvature constant" of a given model in the following sense. There exists a curved surface in Euclidean 3-space, called a pseudosphere, such that the internal geometry of that surface, as derived from the Euclidean geometry of the surrounding space (for example, by taking its geodesics to be its "lines") is identical with hyperbolic geometry (it is in fact isomorphic with a portion of the hyperbolic plane). The constant $1/K$ is related to the so-called Gaussian curvature of that surface. If the surface is enlarged, without change of shape, the value of $1/K$ increases. It is analogous to the radius of a sphere on which one can discuss spherical geometry. In that sense, a two-dimensional geometry is spherical if it has positive curvature, Euclidean if it has zero curvature, and hyperbolic if it has negative curvature. In these cases, the value of the curvature is the same at all points of the space, and there is a continuous spectrum of constant-curvature models from spherical through Euclidean to hyperbolic, but that spectrum contains just three abstract geometries, and they have different axiom systems.

1.3 Comments on Spherical Geometry

Although hyperbolic geometry is our main concern, we wish to point out certain similarities and contrasts with the geometry on a sphere, which we assume to be known. The "lines" on the sphere are the great circles, and the distances are distances along great circle routes. However, when points A and B are given on a great circle, there generally is a short segment from A to B and a longer segment, which "goes the long way round"; hence we shall speak of the length of a segment rather than the "distance" between the points. Most of the axioms in Section 1.1 are valid much of the time, but can fail. Axiom 1 fails if the given points are antipodal points on the sphere, for then there are infinitely many great circles through them. The triangle inequality in Axiom 2 can fail if $|AB|$ means the length of the segment AB, for if that segment covers more than half a great circle, then the segments BC and CA can be so short that $|AB|$ is greater than the sum of $|BC|$ and $|CA|$. Axiom 3 fails because a line (great circle) is not homeomorphic to $\mathbb{R}$, but to a circle, which it in fact is. One can use a coordinates θ going from 0 to 2π along a great circle (assuming that the sphere is of unit radius) such that the length of segment AB is either $|\theta(A) - \theta(B)|$ or 2π minus that quantity. The separation Axiom 4 holds. Axiom 5 on measures of angles holds. Axiom 6, the SAS criterion, holds in all cases, although the "triangle" may look rather peculiar if the side lengths are equal to or greater than π. The parallel axioms are false, because there are no parallels (any two great circles intersect).

An important variation of spherical geometry (often called elliptic geometry) is that in which antipodal points are identified; that is, in which a "point" of the geometry is a pair of antipodal points or, if you wish, is a diameter of the sphere. In this case, Axiom 1 is restored. Any two *distinct* "points" determine a unique line (great circle), but the separation Axiom 4 fails. The two hemispheres bounded by a great circle are now one and the same thing, because their points have been identified.

In either of these geometries, the sum of the angles of a triangle exceeds π by an amount called the *angular excess*. The area of a triangle is equal to a^2 times the angular excess, where a is the radius of the sphere and where angles are expressed in radians.

In Chapter 5 a quantity called the *Gaussian curvature* is defined on a surface on which there is a geometry. It is positive for spherical geometry, zero for Euclidean, and negative for hyperbolic geometry. In general it varies from one point of a surface to another, but it is constant in the cases mentioned (hyperbolic, Euclidean, spherical, elliptic), which are the most general two-dimensional geometries of constant curvature. There is thus a spectrum of such geometries, from hyperbolic through Euclidean to spherical. However, as just noted, the axioms for the spherical case are rather different from those of the other two, so that hyperbolic geometry is closer to Euclidean than to other non-Euclidean geometries.

The discussion of the axioms of three-dimensional hyperbolic geometry is postponed to Chapter 4.

Appendix: The Real Number System

Since we have made use of the real number system $\mathbb{R}$, its axioms are in a sense included in the overall axiom system of the hyperbolic plane. Here we discuss those axioms and some of their consequences, but we omit most of the proofs, which belong more properly in a course in real analysis. The system $\mathbb{R}$ is a complete ordered field. We give the axioms of a field, and we discuss the completeness axiom in some detail. The smallest subfield of $\mathbb{R}$ is isomorphic to the field of real rational numbers. We show that the axiom system is categorical, that is, that any two models of $\mathbb{R}$ are isomorphic, and we present the most important models: the decimal model, the model based on equivalence classes of Cauchy sequences of rational numbers, and the model based on Dedekind cuts. We then discuss briefly the apparent contradiction between the categoricalness and Gödel's incompleteness theorem. We assume a knowledge of the integers and the rational numbers. For anyone interested in the set-theoretical foundation of the integers and rational numbers, we recommend the first four chapters of the book by Herbert B. Enderton, (*Elements of Set Theory* Academic Press, 1977).

A.1 Axioms of an Ordered Field

The real number system $\mathbb{R}$ is a complete ordered field. In this section we state the axioms of an ordered field (not necessarily complete); completeness is discussed in Section A.3 below. Informally speaking, a field is a set of quantities that satisfies all the usual commutative, associative, and distributive laws of arithmetic. An ordered field is a field in which there is an order relation $a < b$, which obeys all the standard arithmetic rules for such an ordering.

Definition 1: An *ordered field* is a set $\mathbb{F}$ of elements $a, b, \ldots, x$ subject to the following rules (axioms):
  (1) $\mathbb{F}$ contains at least two distinct elements. (See "Note" below.)
  (2) *Addition* is defined in $\mathbb{F}$: if a and b are any elements of $\mathbb{F}$, then their sum, denoted by $a + b$, is also in $\mathbb{F}$.

(3) *Multiplication* is also defined in $\mathbb{F}$: if a and b are any elements of $\mathbb{F}$, then their product, denoted by ab, is also in $\mathbb{F}$.

(4) *Commutativity*: $a + b = b + a$ and $ab = ba$.

(5) *Associativity*: $(a + b) + c = a + (b + c)$; $(ab)c = a(bc)$.

(6) *Distributivity*: $a(b + c) = ab + ac$.

(7) *Additive identity*: There is a unique element O such that $a + O = a$ for all a in $\mathbb{F}$.

(8) *Multiplicative identity*: There is a unique element E such that $aE = a$ for all a in $\mathbb{F}$.

(9) *Additive inverse*: For every a there is a unique element x such that $a + x = O$; x is denoted by $-a$. For any a, b, $a + (-b)$ is denoted by $a - b$.

(10) *Multiplicative inverse*: For every $a \neq O$, there is a unique x such that $ax = E$. x is denoted by a^{-1} and ba^{-1} is also denoted by b/a; we say that b/a is the result of *dividing* b by a.

(11) *Law of cancellation*: If $ab = ac$, where $a \neq O$, then $b = c$.

(12) An *order relation* $<$ is defined in $\mathbb{F}$. If a and b are any elements, then just one of the relations $a < b$, $a = b$, $a > b$ (which means $b < a$) holds.

(13) *Sum and product of positive elements*: If $a > O$ and $b > O$, then $a + b > O$ and $ab > O$. (An element greater than O is called *positive*.)

(14) *Order preserved by addition*: If $a < b$, then $a + c < b + c$ for all c.

Note: The set consisting of the single element O satisfies all the axioms (2) through (14) (some of them vacuously), provided that we define $O + O = O$, $OO = O$, and $E = O$. The purpose of axiom (1) is to avoid this trivial case. It will be seen that if $\mathbb{F}$ contains at least two distinct elements and satisfies the axioms, then it contains infinitely many elements; it contains at least a subfield $\mathbb{F}_0$ isomorphic to the field $\mathbb{Q}$ of real rational numbers (see next section).

We assume it is known that the rational number system $\mathbb{Q}$ satisfies all these axioms; we obviously must require that any model of the real number system satisfy them, too. Other laws, just as obvious, can be obtained from the ones given above. For example, because of commutativity (Axiom 3), the rules (6) and (7) can be written as $O + a = a$ and $Ea = a$, and the distributive law (5) can be written as $(b + c)a = ba + ca$ (see the Exercises below).

Mathematical structures in which some, but not all, of these laws hold are discussed in courses on modern algebra; they include groups, semigroups, rings, more general fields (often without ordering), finite fields (necessarily without ordering), integral domains, skew fields, modules, and vector spaces. See *Survey of Modern Algebra* by Birkhoff and Maclane (Macmillan, 1953).

These axioms do not suffice to determine $\mathbb{F}$ completely; for example, some number fields contain the number $\sqrt{2}$ and some do not. However, if the further axiom of completeness is included, as in Section A.3 below, then

$\mathbb{F}$ is determined completely, in the sense of isomorphism (see Section A.4). Under those circumstances, the set of axioms is called *categorical*. Peano's axioms for the natural number system $\mathbb{N}$ are categorical, if the induction axiom is stated in terms of arbitrary subsets of $\mathbb{N}$, whereas the axioms of a group are not. In fact, there are so many different structures (groups) that satisfy all the group axioms that the full classification of finite groups was achieved only recently.

   Terminology: Axioms (1) – (11) determine a *field*; Axioms (1) – (14) determine an *ordered field*. (In books on modern algebra, the term "field" is used more generally; in the terminology of those books, the structures we are concerned with are called "fields of characteristic zero.")

Exercises. Prove the following consequences of the axioms:

1. $a \sum_{j=1}^{n} b_j = \sum_{j=1}^{n} ab_j$. (Use induction on n.)
2. $O + a = a$ for all a.
3. $Ea = a$ for all a.
4. $(b + c)a = ba + ca$.
5. $(-(-a)) = a$.
6. If $a > O$, then $(-a) < O$ and if $a < O$, then $(-a) > O$. *Hint for the first:* If $a > O$, then either of the assumptions $(-a) > O$ and $(-a) = O$ leads to a contradiction.
7. $a < b$ if and only if $b - a > O$.
8. (Transitive law) If $a < b$ and $b < c$, then $a < c$.
9. If $a < b$ and $c > O$, then $ac < bc$.
10. $aO = O$ for all a.
11. $(-a)(-b) = ab$. *Hint:* Multiply the equation $a + (-a) = O$ by b and multiply the equation $b + (-b) = O$ by $(-a)$.
12. $E^2 = E$.
13. $E > O$.

A.2 The Rational Subfield

Let $\mathbb{F}$ be an ordered field. A subset of the set $\mathbb{F}$ that satisfies all the axioms is a *subfield*. Let $\mathbb{F}_0$ be the intersection of all subfields of $\mathbb{F}$, that is, the set of all elements of $\mathbb{F}$, each of which is in every subfield. Clearly, $\mathbb{F}_0$ is a subfield (see Exercise 1 at the end of this section); it is in fact the smallest ordered field contained in $\mathbb{F}$. We wish to describe $\mathbb{F}_0$.

   First, any subfield contains the elements O and E, because it is required to contain elements with their properties, and, according to the axioms, the elements O and E are unique in $\mathbb{F}$. Therefore, $\mathbb{F}_0$ contains them, and it contains all elements that can be obtained from O and E by finitely many additions, subtractions, multiplications, and divisions (where, of course, division by an element equal to O is excluded). For example, $\mathbb{F}_0$ contains the element $(E + E)/(E + E + E)$ (which we shall soon identify with the number $2/3$). We shall now show that $\mathbb{F}_0$ is isomorphic to the field $\mathbb{Q}$ of rational numbers by defining an order-preserving isomorphism ϕ of $\mathbb{Q}$ onto $\mathbb{F}_0$.

We start by defining $\phi(0) = O$, $\phi(1) = E$, and, for any positive integer n, we define $\phi(n) = E + E + \cdots + E$, where there are n terms in the sum. We see that $\phi(n) \neq O$ for any positive integer n, because a sum $E + \cdots + E$ is always positive. Clearly, $\phi(n) + \phi(m)$ is a sum $E + \cdots + E$ of $n + m$ terms; hence $\phi(n + m) = \phi(n) + \phi(m)$. If we multiply out a product $\phi(n)\phi(m) = (E + \cdots + E)(E + \cdots + E)$, where the first factor contains n terms and the second m terms, and we use the equation $E^2 = E$ (Exercise 12 in the preceding section), we see that $\phi(n)\phi(m) = \phi(nm)$. We define ϕ for negative integers by saying that if $n > 0$, $\phi(-n) = -E - E - \cdots - E$, where there are n terms. This leads to the following result, of which the remainder of the proof is left to Exercise 2 below.

Lemma: The mapping $n \mapsto \phi(n)$ is such that, for any integers n, m (positive, negative, or zero), we have
(1) $\phi(n) = \phi(m)$ if and only if $n = m$.
(2) $\phi(n) + \phi(m) = \phi(n + m)$.
(3) $\phi(n)\phi(m) = \phi(nm)$.
(4) If $n < m$, then $\phi(n) < \phi(m)$.

So far, we have considered only the integers in $\mathbb{Q}$. We denote by $\mathbb{Z}$ the set of all integers $\ldots, -2, -1, 0, 1, 2, \ldots$, and we denote by $\phi(\mathbb{Z})$ the set of all elements of $\mathbb{F}$ of the form $\phi(n)$ for n in $\mathbb{Z}$. *Note:* $\mathbb{Z}$ and $\phi(\mathbb{Z})$ are not fields, because they do not have multiplicative inverses. They are in fact ordered *rings*, and ϕ is an isomorphism of the ring $\mathbb{Z}$ onto the ring $\phi(\mathbb{Z})$.

To define $\phi(x)$ for an arbitrary rational number x, we write x in lowest terms as $x = p/q$, where p and q are relatively prime integers and $q > 0$. Since $\phi(q) \neq O$, we can divide by it, and we define $\phi(x) = \phi(p)/\phi(q)$. If this same x is expressed as a fraction m/n not in lowest terms, there is a nonzero integer r such that $m = rp$ and $n = rq$. Then,

$$\frac{\phi(m)}{\phi(n)} = \frac{\phi(rp)}{\phi(rq)} = \frac{\phi(r)\phi(p)}{\phi(r)\phi(q)} = \frac{\phi(p)}{\phi(q)} = \phi(x).$$

That is, $\phi(m/n) = \phi(m)/\phi(n)$, whether m/n is in lowest terms or not. From that it is easy to establish that $\phi(x+y) = \phi(x)+\phi(y)$ and $\phi(xy) = \phi(x)\phi(y)$. That leads to the following theorem.

Theorem: The mapping $x \mapsto \phi(x)$ is an order-preserving isomorphism of the field $\mathbb{Q}$ onto the field $\mathbb{F}_0$.

The remainder of the proof is left to Exercise 3. It must be proved that

$\phi(x) = \phi(y)$ if and only if $x = y$; that all elements of $\mathbb{F}_0$ are included; and that $\phi(x) < \phi(y)$ if and only if $x < y$. Henceforth, we shall speak of the elements of $\mathbb{F}_0$ as "rational numbers." When we refer to a rational number x as an element of $\mathbb{F}$, we mean the corresponding element of $\mathbb{F}_0$. In this sense, every ordered field *contains* $\mathbb{Q}$ as a subfield.

Exercises

1. Prove that the intersection $\mathbb{F}_0$ of all subfields of an ordered field is a set that satisfies all the axioms for ordered fields.
2. Prove the Lemma.
3. Prove the Theorem.

A.3 The Completeness Axiom

Two of the more common statements of the completeness of the real number system $\mathbb{R}$ are (1) any nonempty set in $\mathbb{R}$ that is bounded above has a least upper bound, and (2) every Cauchy sequence in $\mathbb{R}$ has a limit in $\mathbb{R}$. As a further axiom to be applied to an ordered field, the two statements are not quite equivalent because of a further property of $\mathbb{R}$ called Archimedean.

Definition 1: An ordered field is *Archimedean* if for any x in the field there is an integer $n > x$.

Statement (1) above of completeness implies the Archimedean property, while (2) does not. For an example of a non-Archimedean ordered field, see Chapter 9, Section 66 of B. L. van der Waerden, *Modern Algebra* (Frederick Ungar, 1949, published originally in German by Julius Springer, Berlin, 1931). If $\mathbb{F}$ contains an element x that is greater than or equal to any positive integer, then $1/x$ is a positive number less than the reciprocal of any positive integer. Such "infinite" and "infinitesimal" elements play a role in Robinson's theory of "nonstandard analysis." To avoid infinities and infinitesimals we take the first statement above as the completeness axiom.

Definition 2: An ordered field $\mathbb{F}$ is *complete* if every nonempty set S in $\mathbb{F}$ that has an upper bound has a least upper bound in $\mathbb{F}$.

The final axiom of the real number system is then this:

Axiom 15: The real number system $\mathbb{R}$ is a complete ordered field.

It is easy to see that an ordered field that is complete in the sense of Definition 2 is Archimedean. The proof is indirect. If there were an element $x \geq$ every integer, then the set S of all integers would be bounded, hence would have a least upper bound B. Then, if n is any integer, $n+1$ would be less than or equal to B – hence n would be less than or equal to $B-1$; $B-1$ would *also* be a bound of S; and that would contradict the assumption that B is the *least* upper bound.

For an Archimedean ordered field, the two statements of completeness in the first paragraph are equivalent, and there are various other equivalent statements. For example:

(3) A nonempty set that is bounded below has a greatest lower bound.
(4) Every bounded monotonic sequence has a limit.
(5) Every continuous function defined on a bounded closed interval attains a maximum (hence also a minimum).
(6) Every continuous function has the intermediate-value property.

Exercises. In the following, $\mathbb{F}$ is always a complete (hence Archimedean) ordered field, as defined above.

1. Prove that if a is any positive element of $\mathbb{F}$, there are rational numbers x such that $0 < x < a$.
2. Prove that if a in $\mathbb{F}$ is such that $-\varepsilon < a < \varepsilon$ for every rational number $\varepsilon > 0$, then $a = 0$.
3. Prove that if $a < b$ in $\mathbb{F}$, then there is a rational x such that $a < x < b$.
4. Prove that if B is the least upper bound of a set S in $\mathbb{F}$, then for any positive ε in $\mathbb{F}$, there are elements x of S such that $B - \varepsilon < x \leq B$.
5. Prove that a bounded nondecreasing sequence has a limit in $\mathbb{F}$, and so does a bounded nonincreasing sequence.
6. Prove that a Cauchy sequence has a limit in $\mathbb{F}$.

A.4 Categoricalness of the Axioms

In the next section we describe several of the familiar models of the real number system $\mathbb{R}$, including the model based on (generally unending) decimals, the model based on Cauchy sequences of rational numbers, and the model based on Dedekind cuts of the rational number system. It is fairly easy to establish that each of these models satisfies all the axioms of a complete ordered field. Here, we show that all such models are isomorphic, hence can be regarded as intrinsically indistinguishable. Let $\mathbb{F}$ and $\mathbb{F}'$ be two complete ordered fields. We seek a one-to-one mapping ψ of $\mathbb{F}$ onto $\mathbb{F}'$ that preserves both the arithmetic operations and the order; hence it is an isomorphism.

Let $\mathbb{F}_0$ and $\mathbb{F}'_0$ be the rational subfields of $\mathbb{F}$ and $\mathbb{F}'$; according to the preceding section, they are both isomorphic to the field $\mathbb{Q}$ of real rational numbers; hence they are isomorphic to each other. We denote that isomorphism by ϕ, so that for any x in $\mathbb{F}_0$, $\phi(x)$ is the corresponding element of $\mathbb{F}'_0$. The problem is to extend ϕ to an isomorphism ψ of all of $\mathbb{F}$ onto all of $\mathbb{F}'$.

We use the technique of Dedekind cuts. For any x in $\mathbb{F}$ (not necessarily rational) we define two subsets L_x and U_x of $\mathbb{F}_0$, called "lower" and "upper" subsets, as follows:

$$L_x = \{a \text{ in } \mathbb{F}_0 : a < x\}, \qquad U_x = \{a \text{ in } \mathbb{F}_0 : a \geq x\}. \qquad \text{(A.4-1)}$$

The pair L_x, U_x is a *Dedekind cut* of $\mathbb{F}_0$, which means that they are disjoint subsets (no rational element is in both of them), their union is all of $\mathbb{F}_0$, every a in L_x is less than every b in U_x, and finally L_x has no greatest member (if a is any rational element less than x, there is always another rational b such that $a < b < x$). The given x can be characterized as

$$x = LUB(L_x) = GLB(U_x). \qquad \text{(A.4-2)}$$

If x is rational, it is the least element of U_x.

Under the mapping mentioned above of $\mathbb{F}_0$ onto $\mathbb{F}'_0$, L_x and U_x are mapped onto subsets of $\mathbb{F}'_0$ given by

$$\phi(L_x) = \{\phi(a) : a \text{ in } L_x\}, \qquad \phi(U_x) = \{\phi(a) : a \text{ in } U_x\}. \qquad \text{(A.4-3)}$$

Lemma 1: $\phi(L_x)$ and $\phi(U_x)$ constitute a Dedekind cut of $\mathbb{F}'_0$.

The proof is left to the reader as Exercise 1 below. The things that have to be proved are these:
(1) $\phi(L_x)$ and $\phi(U_x)$ are disjoint.
(2) The union of $\phi(L_x)$ and $\phi(U_x)$ is all of $\mathbb{F}'_0$.
(3) Every a' in $\phi(L_x)$ is less than every b' in $\phi(U_x)$.
(4) $\phi(L_x)$ has no greatest member.
    We now define the mapping ψ from $\mathbb{F}$ to $\mathbb{F}'$ by saying that for any x,

$$\psi(x) = LUB(\phi(L_x)). \qquad \text{(A.4-4)}$$

Lemma 2: The mapping ψ is an extension of ϕ and is a one-to-one mapping of $\mathbb{F}$ onto $\mathbb{F}'$ which preserves order.

The proof is left to Exercise 2. One has to prove the following:
(5) If x is rational, then $\psi(x) = \phi(x)$.
(6) If $x < y$, then $\psi(x) < \psi(y)$.
(7) For any x' in $\mathbb{F}'$ there is an x in $\mathbb{F}$ such that $x' = \psi(x)$.

Lemma 3: ψ preserves addition: $\psi(x + y) = \psi(x) + \psi(y)$.

Lemma 4: ψ preserves multiplication: $\psi(xy) = \psi(x)\psi(y)$.

Main Theorem: Any two complete ordered fields are isomorphic.

Exercises

1. Prove Lemma 1.
2. Prove Lemma 2. *Hint for (7):* Make use of the inverse mapping ϕ^{-1}.
3. Prove Lemma 3. *Hint:* For any positive rational δ, there are rational elements a_1, a_2 such that $a_1 < x < a_2$ and $a_2 - a_1 < \delta$. For elements of $\mathbb{F}_0$, ψ is the same as ϕ and preserves order. Derive inequalities

$$-\phi(\delta) < \psi(x) + \psi(y) - \psi(x+y) < \phi(\delta),$$

and show that $\phi(\delta)$ can be regarded as an arbitrary positive element of $\mathbb{F}'_0$, so that Exercise 2 of the preceding section can be used.
4. Prove Lemma 4. *Hint:* Do it first when x and y are both positive.
5. Let a be a positive element of $\mathbb{F}$. Let $x = LUB\{y \text{ in } \mathbb{F}_0 : y^2 < a\}$, and show that $x^2 = a$.

A.5 Some Models of ℝ

The constructions of the models start with the field $\mathbb{Q}$ of rational numbers. We assume a complete knowledge of the properties of rational numbers, including terminating decimals.

(A) *The decimal model.* In this model, each number is represented by a decimal (e.g., $+31.672292\cdots$) containing a $+$ or $-$ sign, followed by finitely many digits to the left of the decimal point and generally infinitely many to the right of it. If the sign is omitted, the $+$ sign is understood. Terminating or "finite" decimals are included by simply filling them out to infinity with zeros, e.g., $31/5 = 6.2000\cdots$, although the zeros and dots are often omitted in writing. The digits to the left of the decimal point constitute an integer, called the *integer part* of the number. Any *prefixed zeros*, that is, zeros to the left of the integer part, can be ignored, so that for example, $008.174\cdots$ means $8.174\cdots$. If the integer part is zero, we indicate it by a single zero digit, so that we write $0.333\cdots$ for $1/3$. If all the other digits are also zero, there is no sign; in that case, we often write that number simply as "0" and omit the decimal part entirely.

The terminating decimals are rational; a general rational number is a decimal that ultimately repeats, for example $3/14 = 0.2\overline{142857}$, which means that the digit combination 142857 is repeated infinitely often on the right.

We recall the ambiguity concerning decimals that end in an infinite string of consecutive nines. For example, if we divide 1 by 3, we get $1/3 = 0.333\cdots$, and if we divide 2 by 3, we get $2/3 = 0.666\cdots$. Adding the two gives $1 = 0.999\cdots$. Hence, we must assume that $1.000\cdots$ and $0.999\cdots$ are the same number. Similarly, we must assume that

$$6.314999\cdots = 6.315000\cdots. \tag{A.5-1}$$

We assume that whenever a decimal ending in all nines results from an arithmetic operation, it is converted to the other form in the manner of the above equation.

Definition: $\mathbb{R}_{\text{dec}}$ is the set of all decimals, as described above, subject to the following rules:
Rule (1). No decimal ends in an infinite string of consecutive nines.
Rule (2). Zeros prefixed to the integer part are omitted, except for a single zero when the integer part of the number is 0.
Rule (3). The zero of $\mathbb{R}_{\text{dec}}$ is $0.000\cdots$, often abbreviated as "0"; it has no $\pm$ sign.

Further rules for the order and the arithmetic operations are described below. *Note*: The "kth decimal place" refers to the kth place after the decimal point, and the "kth digit" refers to the digit in the kth place.

To define the order relation $<$, we first consider the case in which a and b are positive. If the integer part of a is less than that of b, we say that $a < b$. If the integer parts are the same, we compare the digits of a with those of b, starting at the decimal point and working to the right. Suppose that in this comparison we find disagreement for the first time in the kth decimal place. If the kth digit of a is less than that of b, we say that $a < b$. If a is negative and b positive, we write $a < 0 < b$. If a and b are both negative, we say that $a < b$ if $-b < -a$.

We next note one aspect of completeness. If $\{x_n\}$ is a bounded nondecreasing sequence, it has a limit in $\mathbb{R}_{\text{dec}}$. Suppose first that the x_n's are positive; we write each of them in the form $N.d_1 d_2 d_3 \cdots$. It is easy to see that, as $n \to \infty$, the integer part N eventually settles down to a value N^* and does not change after that. Then, for n large enough that $N = N^*$, the digit d_1 settles down to a value d_1^* and does not change after that; then, for members of the sequence such that $d_1 = d_1^*$, d_2 ultimately settles down to a value d_2^*, and so on. We call x^* the number $N^*.d_1^* d_2^* d_3^* \cdots$, modified in the manner of (A.5-1) in case it ends in a string of nines. It is easy to prove that the sequence converges to x^*.

We now define addition and multiplication. First, suppose that a and b are positive. We denote by $(a)_n$ and $(b)_n$ their truncations to n places of decimals, that is, the result of replacing all digits after the nth by zeros. Then, we can form the sum $(a)_n + (b)_n$, because the truncations are ordinary terminating decimals, and we can form also the product $(a)_n(b)_n$; the sequences $(a)_n + (b)_n$, $n \geq 1$, and $(a)_n(b)_n$, $n \geq 1$, are bounded nondecreasing sequences, and we define $a + b$ and ab to be their limits. It is clear what to do if either or both of a and b is negative or zero.

To show that the resulting structure $\mathbb{R}_{\text{dec}}$ satisfies all the axioms of a complete ordered field is now a matter mainly of juggling inequalities; it may be a bit tedious, but it is straightforward.

(B) *The Cauchy sequence model.* We wish the model to be such that, in particular, every Cauchy sequence of rational numbers has a limit in $\mathbb{R}$. In the special case that the limits are also rational, we know that two Cauchy sequences $\{x_n\}$ and $\{y_n\}$ of rational numbers have the same limit if and only if they are *equivalent* sequences, that is, if and only if the sequence $\{x_n - y_n\}$ converges to zero. We therefore define each element of this model of $\mathbb{R}$ to be an equivalence class of Cauchy sequence of rational numbers. For the proof that this model satisfies all the axioms of an ordered field, see any good book on advanced calculus or real analysis. Let us just point out, however, what has to be proved by way of completeness. Let $\{z_n\}$ be a Cauchy sequence of elements (not necessarily rational) of $\mathbb{R}$. Then, each z_n is an equivalence class of Cauchy sequences of rationals, and we have to prove that there is a limit L (another equivalence class of Cauchy sequences of rationals) to which $\{z_n\}$ converges.

(C) *The Dedekind cut model.* Dedekind cuts were defined in the preceding section. A Dedekind cut of the rational number system $\mathbb{Q}$ is a disjoint pair L and U of nonempty subsets of $\mathbb{Q}$ such that each element of L is less than each element of U, such that the union of L and U is all of $\mathbb{Q}$, and so that L has no greatest member. [The last means merely that if there is a rational number that is the least upper bound of L and also the greatest lower bound of U, then that rational number is put into U rather than L. That is purely arbitrary; it could have been done the other way.] Each Dedekind cut (L, U) of $\mathbb{Q}$ is defined to be an element of this model of $\mathbb{R}$, and the set of all of them is made into a field $\mathbb{F}$ by suitably defining the sum and the product of any two elements — see Exercise 2 below. The rational subfield $\mathbb{F}_0$ consists of those cuts (L, U) such that U has a least element, which is necessarily rational, because all the numbers in L and U are rational. The order is defined by saying that $(L, U) < (L', U')$ if L is a proper subset of L'. For completeness, see Exercise 3.

In each of these models, we have added to $\mathbb{Q}$ further elements so as to make the smallest complete ordered field of which $\mathbb{Q}$ is a subfield. That process is called *completing* the field $\mathbb{Q}$. Since the results are isomorphic, it is a matter of taste which method of completing $\mathbb{Q}$ is to be preferred. The completion by decimals may be slightly preferred for advanced calculus, because it is closer to familiar things (decimals), but the method of completion by Cauchy sequences has many other applications in mathematics, for example, to Banach and Hilbert spaces in functional analysis.

Exercises

1. Let $\{x_n\}$ be a Cauchy sequence of elements of $\mathbb{R}_{\text{dec}}$. Show how to construct the limit L, decimal by decimal. *Note:* Two cases can arise; for some such sequences, the kth digit of x_n settles down to a fixed value, as n goes to infinity, for every k. In other cases, there may be at least one decimal place where the digit never settles down, but takes on each of two successive values for infinitely many values of n.

2. In the Dedekind cut model, show how to define the sum and product
    of two cuts so that the laws of arithmetic are satisfied.
3. Show that the Dedekind model is complete. *Hint:* When any collection
    of Dedekind cuts is given, consider the union of the lower parts L and
    the intersection of the upper parts U.

A.6 Categoricalness and the Gödel Incompleteness Theorem

The practical importance of categoricalness is that it permits one to use
any model for proving things about the real number system $\mathbb{R}$. An example
is provided by $\sqrt{2}$. According to Exercise 2 in Section A.4, it follows from
the axioms that in any complete ordered field there is a number x whose
square is 2. However, it is not necessary to go back to the axioms to prove
that, provided one accepts the decimal model $\mathbb{R}_{dec}$. There is a method,
sometimes taught to young students, for extracting square roots by decimals
to any number of decimal places. It then follows from the discussion in the
preceding section that in $\mathbb{R}_{dec}$ there is a number x whose square is 2, and
then it follows from categoricalness that the same is true in any other model.

In a sense, then, a categorical set of axioms determines a mathematical
structure completely. That might seem to contradict the work of Gödel and
Cohen. Gödel proved that if an axiom system is sufficiently rich (in fact, it
only has to be rich enough to imply the existence of the positive integers),
then there are statements (called "undecidable") about the system that
cannot be proved or disproved by use of the axioms. (One may then be able
to prove or disprove a particular statement by introducing further axioms,
but, no matter how many axioms are introduced, there will always be some
such undecidable statements — that is Gödel's incompleteness theorem.)
Since a property of a structure can be described by making a statement, it
would then seem that no set of axioms can determine a structure completely,
that is, can determine all its properties.

To resolve this paradox, we must be more precise as to what constitutes
a *statement* about a structure. The statement can refer to the concepts
contained in or implied by the axioms ("number," "less than," or "interval"
in arithmetic, or "point," "line," or "triangle" in geometry), but it must say
things about those concepts only according to a set of rules of discourse.
Then, any property of the system specified by a proper statement, in this
sense, is determined by the axioms, if they are categorical.

Unfortunately, we cannot specify the rules of discourse completely
(that would take us too far into logic and set theory). We merely describe
an example. Gödel proved that the continuum hypothesis, introduced by
Cantor late in the 19th century, cannot be disproved by the usual axioms of
set theory. Later, Paul Cohen showed that it cannot be proved either. The

hypothesis says that there are no sets of numbers with cardinality between that of the integers (which is the same as that of the rational numbers) and that of the continuum, which is that of $\mathbb{R}$. The statement of the hypothesis is this:

"For every subset S of $\mathbb{R}$, either there is a one-to-one correspondence between S and a set of integers, or there is a one-to-one correspondence between S and $\mathbb{R}$."

This statement is excluded (i.e., is not a proper statement about $\mathbb{R}$), because one of the rules says that the quantifiers ("for every" or "there exists") can refer to specifiable objects, but not to arbitrary subsets of a given set.

Therefore, categoricalness must be interpreted as saying that, within the universe of concepts and relations that can result by the rules of discourse from our axioms, our set of axioms determines $\mathbb{R}$ completely. That is, all properties of $\mathbb{R}$ that can be described within that universe are determined uniquely by the axioms.

Chapter 2

Some Neutral Theorems of Plane Geometry

We discuss some of those theorems of Euclidean plane geometry that are independent of the parallel axiom. They will be needed in the development of hyperbolic geometry. We assume they are more or less known, so that our treatment is not as complete as a full treatment of Euclidean geometry would be. Some of the theorems are weaker than the corresponding Euclidean ones, because the more complete form would require the parallel axiom. Some of them go a little beyond Euclid in that they use the notion of continuity as it appears in calculus. Results in the last two sections go beyond Euclid in that the ideas in them are more recent, as in the Jordan Curve Theorem or the study of isometries.

2.1 Neutral Theorems

A *neutral* theorem is one whose proof does not involve the parallel axiom, either directly or indirectly; such a theorem is valid in both Euclidean and hyperbolic geometry. Most such theorems are found in classical Euclidean geometry, but there are exceptions, namely, (a) theorems that ought to have been stated and proved in Euclid's *Elements*, but were assumed without proof; (b) theorems that, although valid in Euclidean geometry, are not generally mentioned there because they are of interest only in the hyperbolic case; and (c) theorems that involve concepts from analysis. An example of the first kind is the crossbar theorem, discussed below as the first corollary of Theorem 2.3. An example of the second kind is Theorem 2.12 below, which says that the sum of the angles of a triangle is less than or equal to 180°; in the Euclidean case the sum is actually equal to 180°, and the proof is easier than the one given below but requires the Euclidean parallel axiom. Examples of the third kind are theorems (like Theorems 2.2 and 2.3 below) that say that certain functions are continuous, or satisfy a Lipschitz condition, or are differentiable.

We give here some neutral theorems needed in subsequent chapters. The proofs are somewhat sketchy, with some details left to be filled in by the reader, so that this chapter is just a review, which differs from other reviews only in that no parallel axiom is used.

2.2 Alternate Interior Angles Theorem

Theorem 2.1: If a line m cuts across distinct lines ℓ and ℓ' in such a way that the alternate interior angles are equal, as indicated in Fig. 2.2a, then the lines ℓ and ℓ' do not meet.

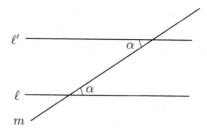

Fig. 2.2a.

The converse of this theorem is false in hyperbolic geometry; the lines may fail to meet even if the alternate interior angles are not equal. An example will be seen in the next chapter, in connection with the so-called angle of parallelism in which one of those angles is a right angle and the other is acute, and still the lines don't meet. The proof of Theorem 2.1 proceeds directly from the axioms, but is indirect. One assumes that the lines *do* meet in a point P, and from that one shows that the two lines would coincide, which contradicts the hypothesis that they are distinct.

Remark: It follows from the equality of vertical angles that the result is also true under the equality of alternate exterior angles, as in the left diagram in Fig. 2.2b, or of angles on the same side, one interior and the other exterior, as in the right diagram.

Proof of Theorem 2.1 (Indirect): Assume, on the contrary, that the lines ℓ and ℓ' meet at a point P, as shown in Fig. 2.2c, even though the angles labeled α are equal. Extend the line AP backward to a point Q such that $|AQ| = |BP|$, and connect points Q and B. Then, by the SAS axiom, the triangles shown are congruent, because AQ is congruent to BP, AB

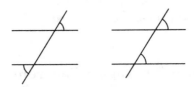

Fig. 2.2b.

is congruent to itself, and the included angles α are equal. It would then follow that the angles labeled β are also equal. By considering the situation at point A, it is seen that $\alpha + \beta = 180°$, and then, by considering point B, it is seen that QB would be an extension of BP; that is, Q would lie on line ℓ, so that ℓ and ℓ' would have two points (P and Q) in common, and that would contradict the hypothesis that they are distinct lines.   □

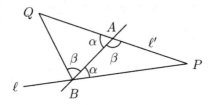

Fig. 2.2c.

2.3 Existence of Perpendiculars; Properties of Certain Functions

Theorem 2.2: Given a line ℓ, a coordinate x on ℓ, and a point Z not on ℓ, then (1) the distance $|ZP|$ for points P on ℓ is a continuous function of $x = x(P)$; that function has a single minimum at a point B and increases to ∞ as $x \to \pm\infty$; (2) the segment ZB is perpendicular to the line ℓ and is unique (there is only one point B such that $ZB \perp \ell$). See Fig. 2.3a.

One says informally that one can always "drop a perpendicular" from Z to the line ℓ. If Z lies on ℓ, the existence of a unique perpendicular to ℓ at Z follows from Axiom 5.

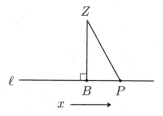

Fig. 2.3a.

In the proof of Theorem 2.2, we need a lemma. It will be proved at this point, but the result will be incorporated in Theorem 2.5 below.

Lemma 2.1 (Isosceles Triangles): If two sides of a triangle are equal, the opposite angles are equal.

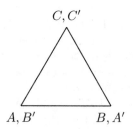

Fig. 2.3b.

Proof: Suppose that, in the triangle $\Delta = ABC$, $|AC| = |BC|$. Denote by $\Delta' = A'B'C'$ the same triangle, but with the vertices labeled differently, as shown in Fig. 2.3b. Then $|AC| = |A'C'|$, $|BC| = |B'C'|$, and $\angle C = \angle C'$, hence the triangles Δ and Δ' are congruent, hence the angles $\angle A$ and $\angle A'$ are congruent, hence $\angle A$ and $\angle B$ are congruent, as required.   □

Proof of Theorem 2.2: If Q is another point on ℓ and $\Delta x = |PQ|$, then, by the triangle inequality, we have

$$|ZP| - \Delta x \leq |ZQ| \leq |ZP| + \Delta x.$$

Therefore, $|ZQ| \to |ZP|$ as $\Delta x \to 0$; hence $|ZP|$ is a continuous function of $x = x(P)$. We now show that $|ZP| \to \infty$, as $x \to \pm\infty$. If A is any fixed point of ℓ, and if P is a point such that $|PA| = C|ZA|$ where $C > 1$, then, by the triangle inequality again, we have

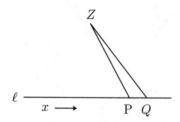

Fig. 2.3c.

$$|ZP| \geq |PA| - |ZA| = (C - 1)|ZA|,$$

which $\to \infty$ as $C \to \infty$. Also, $|ZP|$ is nonnegative; hence, as a function of x, it is continuous and bounded below and becomes infinite as $x \to \pm\infty$ in either direction. It follows that $|ZP|$ takes on a minimum value s_0 for some $x = x_0$, and it follows that if s is any value greater than s_0, $|ZP|$ takes on the value s at least once on either side of x_0. That is, there are points P_1 and P_2 on ℓ such that the triangle ZP_1P_2 is isosceles, so the angles α shown in Fig. 2.3d are equal. Then, if B is the midpoint of P_1P_2, the triangles ZP_1B and ZP_2B are congruent. Since they are back-to-back, they are right triangles, hence ZB is perpendicular to ℓ. To prove uniqueness of the point B, we note that if there were two points B and B' such that ZB and ZB' were both perpendicular to ℓ, then, by Theorem 2.1 on alternate interior angles, the lines ZB and ZB' would not meet — but they do in fact meet at the point Z. We see also that the function $|ZP|$ increases strictly with respect to x for *all* x greater than x_0 and decreases for all $x < x_0$, for otherwise there would be another pair of points the same distance from Z, and on the same side of the point whose coordinate is x_0, hence there would be another point, say B'', such that ZB'' is perpendicular to ℓ. We have seen that there can be no such point distinct from B. □

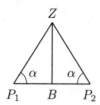

Fig. 2.3d.

Corollary 1: The hypotenuse of a right triangle is longer than either of the other two sides.

Proof: Let ABC be a triangle with right angle at B. Denote by ℓ the line containing BC and relabel the points A, C as Z, P. Then, Theorem 2.2 holds (see Figure 2.3a), hence AB is less than AC. By interchange of A and C, CB is less than AB.   □

Corollary 2: For any points A, B, C, $|AC| = |AB| + |BC|$ if and only if the three points are on a line and B lies between A and C.

Sketch of the Proof: The equation is clearly true if B is in the segment AC and clearly false if B is on the line of A and C but outside the segment AC. If B is not on the line, drop a perpendicular from B to the line, and let D be the foot of the perpendicular. Then AB and BC are the hypotenuses of right triangles, hence are longer than AD and DC.

Definition: The minimum of $|AP|$, as P varies on the line m, is called the *distance of the point A from the line m* and is often denoted by $d(A, m)$.

Corollary 3: The distance $d(A, m)$ is a continuous function of the point A. That is, for any $\varepsilon > 0$ there is a $\delta > 0$ such that if the distance $|AA'|$ is less than δ, then $|d(A', m) - d(A, m)|$ is less than ε. In fact, we can take $\delta = \varepsilon$.

Proof: Let $|A'A| < \varepsilon$. By the triangle inequality, for any P on m, we have $|A'P| \leq |AP| + |A'A| < |AP| + \varepsilon$. Similarly, $|AP| \lessgtr |A'P| + \varepsilon$. Let B be the foot of the perpendicular dropped from A, and B' the foot of the perpendicular dropped from A'. Then, since $|A'B'|$ is the minimum of $|A'P|$, $|A'B'|$ is less than or equal to $|A'P|$ for any P, hence is less than or equal to $|A'B|$ in particular. Therefore, $|A'B'| < |AB| + \varepsilon$, and similarly $|AB| < |A'B'| + \varepsilon$, but $|A'B'|$ is $d(A', m)$ and $|AB|$ is $d(A, m)$, and the continuity follows.   □

The next theorem deals with the dependence on x of an angle.

Theorem 2.3: Given a line ℓ with coordinate x and a point Z not on ℓ and points A and B on ℓ such that $x(A) < x(B)$; then, for points P between A and B, the angle $\alpha = \angle AZP$ is a continuous increasing function of $x = x(P)$, for $x(A) < x < x(B)$. See Fig. 2.3e.

Proof: That the function $\alpha = \alpha(x)$ is continuous follows from part (e) of Axiom 5. To show that it is increasing, let P' be a point to the right of P but between P and B. Then, by the definition of interior, the ray $\overrightarrow{ZP}$ is in the interior of the angle $\angle AZP'$, and it follows that the measure of the angle $\angle AZP'$ is greater than that of $\angle AZP$, according to part (c) of that axiom.   □

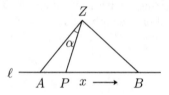

Fig. 2.3e.

Corollary 1 (Crossbar Theorem): Given a triangle ABC and a ray $\overrightarrow{AD}$ in the interior of the angle $\angle CAB$, then the ray intersects the side BC. See Fig. 2.3f.

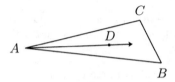

Fig. 2.3f.

Remark: It was discovered in the nineteenth century that Euclid had assumed this result without proof. That there is something that needs proof is indicated by the fact that in the hyperbolic plane it is possible to find three rays $\overrightarrow{h}, \overrightarrow{j}, \overrightarrow{k}$ and a line ℓ such that the ray $\overrightarrow{j}$ and the entire line ℓ are in the interior of the angle $\angle \overrightarrow{h}, \overrightarrow{k}$, but the ray $\overrightarrow{j}$ never intersects the line ℓ, as indicated roughly in Fig. 2.3g. After we discuss the angle of parallelism in Section 3.1, it will be easy to see that such configurations exist. In the Crossbar Theorem it is not sufficient to have a *line* in the interior of that angle; one must have a "crossbar," namely, a segment BC with one end on each of the rays that bound the angle.

Proof: Let θ be the angle $\angle \overrightarrow{AD}, \overrightarrow{AB}$ in Figs. 2.3f and 2.3h. According to part (c) of Axiom 5, θ is between 0 and $\angle BAC$. Let x be a coordinate on BC increasing from B to C as in Fig. 2.3h. For any P on BC with coordinate x, let $\alpha = \alpha(x)$ be the angle $\angle BAP$. Since $\alpha(x)$ is continuous and increasing, there is a value $x = x_0$ such that $\alpha(x_0) = \theta$. Call P_0 the point on BC with coordinate x_0. Then the rays $\overrightarrow{AD}$ and $\overrightarrow{AP_0}$ make the same angle θ with $\overrightarrow{AB}$, hence they are the same ray according to the one-to-one relation between

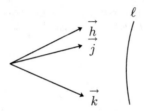

Fig. 2.3g.

rays and angles given by part (d) of Axiom 5. That is, the ray $\overrightarrow{AD}$ and the segment BC intersect in the point P_0, as required.   □

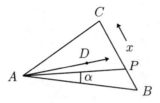

Fig. 2.3h.

Corollary 2 (Pasch's Theorem): If a ray enters a triangle through one side but does not pass through the opposite vertex, then it intersects one (only one) of the other sides.

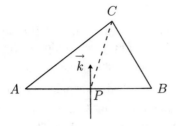

Fig. 2.3i.

Proof: Suppose $\overrightarrow{k}$ intersects side AB at point P, as shown in Fig. 2.3i. According to part (d) of Axiom 5, the angle $\angle \overrightarrow{k}, \overrightarrow{PB}$ is either less than or greater than $\angle CPB$. In the first case, by the crossbar theorem, it intersects side BC of the triangle PCB; in the second case, it intersects side AC of the triangle PCA.   □

2.4 The Exterior Angle Theorem and Its Consequences

Theorem 2.4: An exterior angle of a triangle is greater than each of the opposite interior angles. That is, in Fig. 2.4a, the angle η is greater than each of the angles α and β. Equivalently, the sum of any two angles of a triangle is less than $180°$.

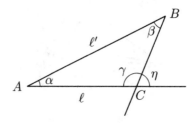

Fig. 2.4a.

Note: In Section 2.7 below, we shall see that the sum of the *three* angles of a triangle cannot exceed $180°$. In the proof of that result, we shall need as a preliminary fact that the sum of any *two* of its angles cannot exceed $180°$.

Proof: We prove first that η is greater than β. If η were *equal* to β, then the lines ℓ and ℓ' would not meet, and we should not have a triangle. Assume, then, that β is greater than η. Construct a ray $\overrightarrow{k}$ from the point B in the interior of the angle at B at an angle η with respect to $\overrightarrow{BC}$, as in Fig. 2.4b. By the crossbar theorem, $\overrightarrow{k}$ would intersect the side AC of the triangle ABC, and that would again contradict the alternate interior angles theorem.   □

Theorem 2.5: If a, b are the lengths of two sides of a triangle and α, β are the angles opposite those sides, then $a = b$ if and only if $\alpha = \beta$, and $a < b$ if and only if $\alpha < \beta$.

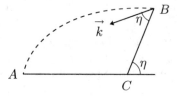

Fig. 2.4b.

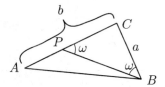

Fig. 2.4c.

Proof: Let ABC be the triangle, with $a = |BC|$ and $b = |AC|$. The angles opposite these sides are $\alpha = \angle CAB$ and $\beta = \angle ABC$. Now assume first that $a < b$, as shown in Fig. 2.4c. Choose point P on AC so that $|PC| = |BC| = a$. Then the triangle PCB is isosceles, so that the angles labeled ω are equal, by Lemma 2.1. Then $\angle ABC > \omega$ by Axiom 5, and $\omega > \angle CAB$ by the exterior-angle theorem (Theorem 2.4) as applied to the triangle APB. Therefore $\alpha < \beta$, as required. Next, by interchanging a and b, we see that if $a > b$, then $\alpha > \beta$. Hence, α cannot be equal to β unless $a = b$. According to Lemma 2.1, if $a = b$, then α *is* equal to β, and this completes the proof. $\square$

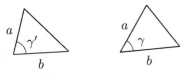

Fig. 2.4d.

Theorem 2.6: If two sides of a first triangle are congruent respectively to two sides of a second triangle, while the included angles are unequal, then the greater third side is opposite the greater angle, as in Fig. 2.4d.

Proof: Construct a copy of the second triangle (the one with the smaller included angle) on the base CA of the first, as in Figs. 2.4e and 2.4f. In either figure, ABC is the triangle with the smaller included angle at C, and $AB'C$ is the one with the larger included angle; also, $|CB| = |CB'| = a$, while $|CA| = b$. Depending on the relative sizes of the two angles at A (not known in advance), there are three possibilities:

(1) B lies on the line of AB'.
(2) B is in the interior of the angle $\angle CAB'$, as in Fig. 2.4e.
(3) B' is in the interior of the angle $\angle CAB$, as in Fig. 2.4f.

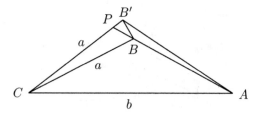

Fig. 2.4e.

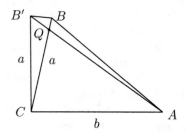

Fig. 2.4f.

In any case, B is in the interior of the angle $\angle B'CA$. In case (1), since A and B are on the same side of CB', AB is a subsegment of AB', hence is shorter, as required. In case (2), the ray $\overrightarrow{AB}$ intersects CB' in a point P, and B lies between A and P, as shown in Fig. 2.4e. Therefore, by combining other facts with two uses of the exterior angle theorem (Theorem 2.4), we have

$$\angle B'BA > \angle PB'B = \angle CBB' > \angle PBB' > \angle BB'A.$$

The first step in this formula is the result of applying the exterior angle theorem to the small triangle $PB'B$; the next step follows from the fact that the triangle CBB' is isosceles; the next step comes from Axiom 5, and the last step follows from applying the exterior angle theorem to the triangle $B'BA$. Then by the preceding theorem (Theorem 2.5), $|B'A| > |BA|$, as required. In case (3), the ray $\overrightarrow{AB'}$ intersects BC in a point Q; since B is in the interior of $\angle B'CA$, Q is also, hence Q is between A and B', as in Fig. 2.4f. Then, we have

$$\angle B'BA > \angle B'BC = \angle BB'C > \angle BB'A.$$

Here, the first and third steps come from Axiom 5 and the second step from the fact that the triangle CBB' is isosceles. Then, by the preceding theorem, $|B'A| > |BA|$, as required.   □

2.5 Congruence Criteria for Triangles

The SAS criterion was Axiom 6. We now discuss the other familiar criteria, SSS, ASA, and SAA. In the hyperbolic plane, there is a fifth criterion, AAA, equality of the angles of a first triangle with the corresponding angles of a second one. In the Euclidean case, that condition gives only *similarity* of the triangles, that is, equality of the ratios of the side lengths. In the hyperbolic plane, it gives also equality of the lengths themselves, but the proof of that will have to wait until later.

Theorem 2.7 (SSS Criterion): If the three sides of a first triangle are congruent respectively to the three sides of a second, then the triangles are congruent.

Proof: We note first that, according to Theorem 2.6, the function $\phi(a, b, \gamma)$ is an increasing function of γ, for fixed a, b. If two triangles had the same values of a, b, c, but different values of γ, then $\phi(a, b, \gamma_1)$ would be equal to $\phi(a, b, \gamma_2)$, for $\gamma_1 \neq \gamma_2$, and that would contradict the increasing nature of the function. We conclude that the triangles have the same value of γ and hence that they are congruent by the SAS criterion.   □

We sometimes write the inverse of the equation $c = \phi(a, b, \gamma)$ as $\gamma = \psi(a, b, c)$, so that $\psi(a, b, c)$ is the angle opposite to the side c, and $\psi(a, b, c) = \psi(b, a, c)$.

Theorem 2.8 (ASA Criterion): If two angles and the included side of a first triangle are congruent respectively to two angles and the included side of a second triangle, then the triangles are congruent.

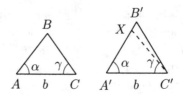

Fig. 2.5a.

Proof: Let the two triangles be ABC and $A'B'C'$, as shown in Fig. 2.5a, where the angles labeled α are congruent, the angles labeled γ are congruent, and the sides labeled b are congruent. On the side $A'B'$ lay off a distance $A'X$ equal to AB. (It will turn out that the point X coincides with B', but we don't assume that in advance.) By the SAS axiom, the triangles ABC and $A'XC'$ are congruent, hence the angles $\angle A'C'X$ and $\angle A'C'B'$ are both equal to γ, hence the line $C'X$ coincides with the line $C'B'$, hence the given triangles are congruent.   □

Theorem 2.9 (SAA Criterion): If a side and two angles, one of them opposite that side, of a first triangle are congruent respectively to a side and two angles, one of them opposite that side, of a second triangle, then the triangles are congruent.

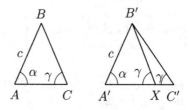

Fig. 2.5b.

Proof: Let the two triangles be ABC and $A'B'C'$, as shown in Fig. 2.5b, where the sides labelled c are congruent, the angles labelled α are congruent, and the angles labelled γ at the vertices C and C' are congruent. On the side $A'C'$ lay off a segment $A'X$ equal in length to AC. Then, by the SAS criterion, the triangles ABC and $A'B'X$ are congruent, hence the two angles labeled γ in the second drawing are equal. By the alternate interior angle theorem (See Remark after Theorem 2.1), if X did not coincide with C', then the lines $B'X$ and $B'C'$ would not meet, and that would contradict the fact that they do meet at B'.   □

Note that there is no angle-side-side criterion. As indicated in Fig. 2.5c, for a given angle $\angle A$ and given lengths $|AB|, |BC|$, there may be two possible locations C_1 and C_2 for the third vertex. However, see the corollary to Theorem 2.11 below for a special case.

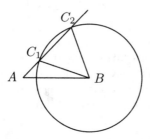

Fig. 2.5c.

We now consider the continuity of the triangle functions. First, a lemma:

Lemma 2.2 (Thin Isosceles Triangles): $\phi(r, r, \gamma) \to 0$ as $\gamma \to 0$ for fixed r; in fact, there is a quantity $h(r)$ such that $\phi(r, r, \gamma) < h(r)\gamma$.

For a proof, see Exercises 1 and 2, below. Now recall that the function $\phi(a, b, \gamma)$, introduced in connection with Axiom 6, gives the third side c of a triangle in terms of the first two sides a, b and included angle γ. Consider two values of γ, called γ_1 and γ_2; in Fig. 2.5d, $\gamma_1 = \angle QAB$ and $\gamma_2 = \angle PAB$.

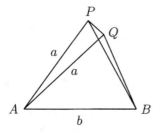

Fig. 2.5d.

By the triangle inequality, we have

$$\left|\phi(a, b, \gamma_2) - \phi(a, b, \gamma_1)\right| = \left||PB| - |QB|\right| \le |PQ| < h(a)(\gamma_2 - \gamma_1),$$

where $h(a)$ is defined in Exercise 1. We see that $\phi(a, b, \gamma)$ is Lipschitz continuous in γ for fixed a and b. If two triangles have the same value of γ and the same value of a or b, with the other varying, we have, again by the triangle inequality, that

$$|\phi(a_2, b, \gamma) - \phi(a_1, b, \gamma)| \leq |a_2 - a_1|,$$
$$|\phi(a, b_2, \gamma) - \phi(a, b_1, \gamma)| \leq |b_2 - b_1|.$$

We see that $\phi(a, b, \gamma)$ is Lipschitz continuous in each of its variables. For the domain of this function, we have assumed that a, b, γ are in open intervals $(0, \infty), (0, \pi)$. The investigation of limiting cases is left to the reader, as is the investigation of the other criterion functions. The conclusion is the following:

Theorem 2.10: In triangle congruences, the remaining sides and angles depend continuously on the given sides and angles.

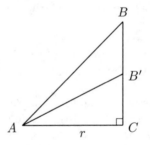

Fig. 2.5e.

Exercises

1. First consider thin *right* triangles by starting with triangle ABC as in Fig. 2.5e, with $|AC| = |BC| = r$, $\angle C = 90°$, and $\angle A = \angle B = \alpha$; α depends only on r. Bisect the angle $\angle A$, letting B' be the point between B and C where the bisector meets the opposite side. Then, by a suitable construction, by use of one of the congruence criteria, and by Theorem 2.2, show that $|B'C|$ is less than half of $|BC| = r$. Then bisect the angle at A of the triangle $AB'C$, and so on, thus producing right triangles with angle at A equal to $2^{-n}\alpha$, $n = 1, 2, \ldots$. Denote the length of the side opposite to A by $\tau(r, 2^{-n}\alpha)$, so that

$$\tau(r, 2^{-n}\alpha) < 2^{-n}r = (r/\alpha)2^{-n}\alpha.$$

For any angle γ less than α choose n so that $2^{-n-1}\alpha < \gamma \leq 2^{-n}\alpha$, and show that $\tau(r,\gamma) < 2r/\alpha \; 2^{-n-1}\alpha < 2r/\alpha \; \gamma = h(r)\gamma$, where $h(r)$ stands for $2r/\alpha$.

2. By a further construction and further use of Theorem 2.2, show that the third side of an isosceles triangle with two sides of length r and included angle γ is less than $\tau(r,\gamma)$, and conclude that $\phi(r,r,\gamma) < h(r)\gamma$, as claimed in Lemma 2.2. *Hint:* Use one of the congruence criteria to show that the bisector of the angle γ is perpendicular to the opposite side.

3. Show the Lipschitz continuity of the right triangle functions defined in terms of Fig. 2.5f as

$$\sigma(c,\alpha) = a, \qquad \kappa(c,\alpha) = b.$$

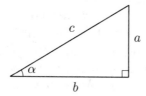

Fig. 2.5f.

4. Show that the functions σ and κ are meaningful for the limiting values $\alpha = 0°$ and $\alpha = 90°$, and give their limiting values for those angles.

2.6 Intersections of Lines and Circles

The distance from a point to a line (if not zero) is defined as the length of the perpendicular ZB of Theorem 2.2. If the point is on the line, the distance is of course taken as zero. The *circle of radius r (> 0) centered at P* is defined (as expected) to be the set of points Q such that $|PQ| = r$. A point Q is *inside* the circle if $|PQ| < r$ and *outside* the circle if $|PQ| > r$. A line is *tangent* to a circle if it has exactly one point in common with the circle, that is, if it has at least one point in common with the circle and is perpendicular to the line through that common point and the center of the circle. There is a tangent at every point of a circle. Two distinct circles are *tangent* if they intersect at a single point. Then, according to Exercise 2 below, they have a common tangent line at that point.

Theorem 2.11: If the distance from the center of a circle to a line is less than the radius of the circle, the line intersects the circle in just two points. If the distance is equal to the radius, there is just one point of intersection, and the line is tangent to the circle. If the distance is greater than the radius, there is no intersection. If one point of a first circle lies inside a second circle while another point of the first circle lies outside, then the two circles meet in just two points; in that case, the distance between the centers is between the difference and the sum of the two radii. If the distance between the centers is equal to the difference or the sum of the radii, the circles are tangent; if the distance between the centers is greater than the sum of the radii or less than the difference, the circles do not intersect.

Proof: The first three statements of the theorem follow from Theorem 2.2, according to which the distance $|AP|$ from a fixed point A to a point P on a line is a continuous function of the coordinate x of P on the line, has a single minimum at a point B, which is the foot of the perpendicular dropped from A to the line, and increases to ∞ as x goes to either $+\infty$ or $-\infty$. Now consider circles of radii a and c with centers at C and A, respectively, as in Fig. 2.6a. If they intersect, say at a point P, then $c = \phi(a, b, \gamma)$, where γ is the angle between $\overrightarrow{CA}$ and $\overrightarrow{CP}$, and b is the distance $|CA|$ between the centers. Hence, the problem of intersection is to find solutions of the equation $c = \phi(a, b, \gamma)$. (We assume that $b \neq 0$, for if the circles are concentric, they are disjoint unless $a = c$, in which case they coincide.) According to Theorems 2.6 and 2.10, $\phi(a, b, \gamma)$ is a continuous function of γ, for fixed a and b, and increases, as γ increases from 0 to π; its limiting values are $\phi(a, b, 0) = |a - b|$ and $\phi(a, b, \pi) = a + b$. Hence, if $c = |a - b|$ or $c = a + b$, there is just one point of intersection, at $\gamma = 0$ or $\gamma = \pi$, respectively, hence on the line of AC. We now assume that $c \geq a$ (if $c < a$, we interchange the circles). The circles are tangent if either

(1) $\gamma = 0$, $b = c + a$ (external tangency), or
(2) $\gamma = \pi$, $b = c - a$ (internal tangency).

If $b < c - a$ or $b > c + a$, the equation has no solution (as also follows from the triangle inequality). If $c - a < b < c + a$, the equation has one solution γ, and there is a point of intersection P in each of the half-planes bounded by the line through the centers C and A. Since the circles intersect or not as described in the theorem when the corresponding equalities or inequalities hold, this completes the proof.   □

In connection with Fig. 2.5c, we noted that there is *in general* no ASS criterion. However, if the angle is 90°, there *is* such a criterion.

Corollary (Right Angle Side Side criterion for congruence of right triangles): If triangles ABC and $A'B'C'$ are such that $\angle C = \angle C' = 90°$, $|CA| = |C'A'|$, and $|AB| = |A'B'|$, as in Fig. 2.6b, then $|BC| = |B'C'|$, hence the triangles are congruent.

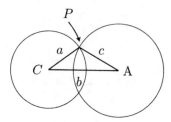

Fig. 2.6a.

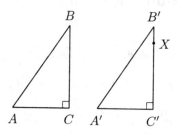

Fig. 2.6b.

Proof: Suppose that $|BC| \le |B'C'|$. Choose a point X on the segment $B'C'$ such that $|C'X| = |CB|$. Then, by the SAS criterion, triangles ABC and $A'XC'$ are congruent, so that $|AB| = |A'X|$. Hence X and B' lie on a circle of radius $|AB|$ about the point A'. Since there is only one point of intersection of the circle and the line $B'C'$ on each side of the line $A'C'$, we see that X and B' coincide, so that $|B'C'| = |XC'| = |BC|$, as required. □

Exercises

1. In Fig. 2.6c assume that the angles labeled α are equal and the sides labeled b are equal. Prove or disprove: It then follows that the two smaller triangles are congruent.

2. Show that if two circles have a single point of intersection, they have a common tangent line at that point.

Fig. 2.6c.

2.7 The Angle Sum of a Triangle

Theorem 2.12: The sum of the three interior angles of a triangle is $\leq 180°$.

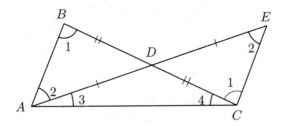

Fig. 2.7a.

Remark: This theorem is usually attributed to Saccheri and Legendre. It is one of the neutral theorems, since it holds for both Euclidean and hyperbolic geometry, but it is not usually regarded as relevant to the Euclidean case, since there the sum of the angles is actually equal to 180°. We give a standard proof. (The theorem does not hold on a sphere.)

Proof (Indirect): Assume that the sum of the angles of the triangle ABC in Fig. 2.7a is $180° + \delta$, where δ is a positive quantity. From this we shall derive a contradiction. Construct the figure shown. D is the midpoint of BC, AD is extended an equal distance to E, as shown. Then, by the SAS criterion and the equality of vertical angles, the triangles ABD and ECD are congruent, hence the two angles labeled 1 are congruent and those labeled 2 are congruent. The original triangle ABC and the triangle AEC both have angle sums equal to $\angle 1 + \angle 2 + \angle 3 + \angle 4$. Angle $\angle A$ is equal to $\angle 2 + \angle 3$, hence the new triangle AEC has one angle (either at A or at E) less than or equal to one-half of the angle $\angle A$ of the original triangle.

By repeating the procedure we get a sequence of triangles, all having angle sum $180° + \delta$ and with one angle less than or equal to $\frac{1}{4}, \frac{1}{8}, \ldots$ times the angle $\angle A$ of the original triangle, hence eventually less than δ. Then the sum of the other two angles would exceed $180°$, and that would contradict the exterior angle theorem (Theorem 2.4).   □

2.8 Quadrilaterals

A *Lambert quadrilateral* is a quadrilateral with three right angles, as in Fig. 2.8a. (In the Euclidean case, the fourth angle is also a right angle, and the quadrilateral is a rectangle.)

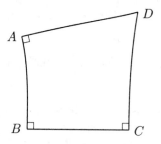

Fig. 2.8a.

Theorem 2.13: In a Lambert quadrilateral, the fourth interior angle is $\leq 90°$ and either side adjacent to that angle is at least as long as the opposite side.

Proof: Let the vertices be A, B, C, D, with right angles at A, B, C, as in Fig. 2.8a. We must prove that the angle at D is $\leq 90°$ and also that $|AB| \leq |DC|$ and $|BC| \leq |AD|$. Draw the diagonal BD. The sum of the six angles of the two resulting triangles is equal to the sum of the four angles of the quadrilateral. By Theorem 2.12 applied to those triangles, we see that the sum of the four angles of the quadrilateral is $\leq 360°$, hence the angle at D is $\leq 90°$. To prove the second part, let E be the midpoint of BC and construct the perpendicular to BC at E, and let F be the intersection of that perpendicular with AD, as in Fig. 2.8b. (That perpendicular cannot intersect either AB or CD, for to do so would create a triangle that would violate Theorem 2.12; hence, by a variant of the crossbar theorem it must intersect AD.) Now construct the quadrilateral $A'B'C'D'$ that results from reflecting the points A, B, C, D in the line EF, as shown in Fig. 2.8b.

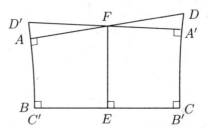

Fig. 2.8b.

Since the segments BE and EC are congruent and the angles at B and C are congruent, the ray $\overrightarrow{B'A'}$ lies along $\overrightarrow{CD}$. Since the angle at A' is $90°$, the point A' cannot lie above D, for if it did, the triangle FDA' would have angle sum greater than $180°$. Therefore $|AB| = |A'B'| \leq |CD|$; the proof that $|BC| \leq |AD|$ is similar. □

An application of these ideas that will be useful in the next chapter is the following:

Theorem 2.14: Consider isosceles triangles all having the same angle α but variable length a of the sides adjacent to α, as in Fig. 2.8c. Then, the length c of the third side is an increasing function of a. Indeed, if a is doubled, c is also at least doubled.

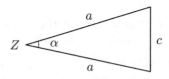

Fig. 2.8c.

Note: This theorem is valid in both the Euclidean and the hyperbolic planes, but is not true on a sphere, where c increases with a until a becomes equal to one-quarter of a great circle and then decreases from then on, becoming zero when the point antipodal to Z is reached.

Proof: In Fig. 2.8d, let ZAB and $ZA'B'$ be isosceles triangles, and suppose that $|ZA'| > |ZA|$. C is the midpoint of AB, and C' the midpoint of $A'B'$.

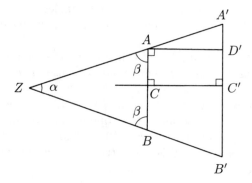

Fig. 2.8d.

We erect the perpendicular AD' to AB at A, as shown. The angles β are less than $90°$, because the angle sum of the triangle ZAB is not more than $180°$, and $\alpha > 0$. Therefore, AD' is in the interior of $\angle A'AC$, so that D' is in the segment $A'C'$, hence $|D'C'| < |A'C'|$. But $ACC'D'$ is a Lambert quadrilateral, so $|D'C'| \geq |AC|$. The fact that c increases with a follows easily. Now suppose $|AA'| = |ZA|$. If $\gamma = \angle D'AA'$, we get $2\gamma \geq \alpha$ because of the inequality on $2\beta + \alpha$ and the equation relating γ and β. In an isosceles triangle with angle 2γ between sides of length a, the third side has length at least c, so the distance from A' to the line through A and D' is at least $c/2$. Hence $|A'D'| \geq c/2$, so $|A'C'| \geq 2|AC|$, as required.   □

Question: How does this proof fail on a sphere?

Corollary: If we consider a set of isosceles triangles with a fixed value of c and variable angle α, then $\alpha \to 0$ as the length $a \to \infty$. That is, in terms of the function $\psi(a, b, c)$ mentioned in Section 2.5, which gives the angle opposite to the side c, the function $\psi(r, r, c)$ decreases with increasing r and fixed c, and $\psi(r, r, c) \to 0$ as $r \to \infty$.

Exercises

1. Show that each of the following criteria implies the congruence of quadrilaterals $ABCD$ and $A'B'C'D'$:

    (SASAS): $|AB| = |A'B'|$, $\angle B = \angle B'$, $|BC| = |B'C'|$, $\angle C = \angle C'$, $|CD| = |C'D'|$.

    (ASASA): $\angle A = \angle A'$, $|AB| = |A'B'|$, $\angle B = \angle B'$, $|BC| = |B'C'|$, $\angle C = \angle C'$.

    (SASAA): $|AB| = |A'B'|$, $\angle B = \angle B'$, $|BC| = |B'C'|$, $\angle C = \angle C'$, $\angle D = \angle D'$.

    Are there other such criteria?

2. A Saccheri quadrilateral is one with right angles at B and C and vertical sides AB and CD of equal length. Prove that the angles at A and D are equal.

3. For the Saccheri quadrilateral in the preceding exercise, show that the perpendicular bisector of the lower side BC also bisects the upper side AD perpendicularly, and conversely.

4. Show that if directed lines ℓ and m intersect, then the distance from a point P on ℓ to m (obtained by dropping a perpendicular) increases with increasing coordinate $x = x(P)$ on ℓ for all x past the point of intersection. (Note that this is not true for great circles on a sphere.) *Hint:* In Fig. 2.8e the angle α is obtuse. (Why?)

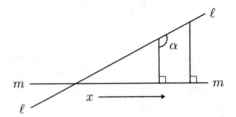

Fig. 2.8e.

2.9 Polygons

The notion of polygon is neutral, and this section provides neutral proofs that it is possible to identify an interior and an exterior of a polygon and that the interior can be divided into triangles.

Definition: A *(parametrized) curve* in the plane $\mathbb{P}$ is a continuous function, C, from a closed interval, $[a, b]$, in $\mathbb{R}$ to $\mathbb{P}$. The curve C is *simple* if it is one-to-one, *closed* if $C(a) = C(b)$, and is called a *simple closed curve* if $C(a) = C(b)$, but otherwise $s \neq t$ implies $C(s) \neq C(t)$. A curve C on $[a, b]$ is *linear* if there is a line ℓ and a coordinate function x on ℓ such that $a \leq t \leq b$ implies $C(t) \in \ell$ and $x(C(t)) = t$. A curve C is called *polygonal* if its domain $[a, b]$ contains a finite number of points, $a = t_0 < t_1 < \cdots < t_n = b$ such that $C|[t_{i-1}, t_i]$ is linear for $i = 1, \ldots, n$. A simple closed polygonal curve will be called a *polygon*, as will the range of the curve, and the points $C(t_i)$, $i = 1, \ldots, n$, will be called its *vertices*. The segments joining successive vertices will be called *edges*. A polygon is of *order n* if it has n vertices.

A polygon of order 3 is a triangle, a polygon of order 4 is a quadrilateral, and so on. Observe that a polygon is the image of a continuous function on a compact set in $\mathbb{R}$ so it is a compact set in $\mathbb{P}$. The reader is encouraged to supply the proof of the following lemma, to help clarify the concept of a linear curve.

Lemma: A curve $C : [a, b] \to \mathbb{P}$ is linear if and only if it is isometric, that is, $a \le s \le t \le b$ implies that $|C(s)C(t)| = t - s$.

For triangles, the interior is easy to define, as we did in Chapter 1, but for more general polygons more work is required to make the notion rigorous. (The intuitive idea is clear enough.) Still more work is needed to handle general simple closed curves. The standard tool is the Jordan Curve Theorem, which says that a simple closed curve divides the plane into two parts, one of which is bounded, and we call the bounded part the interior of the curve.

We will prove the Jordan Curve Theorem for the special case of polygons in order to provide the needed rigor. The proof we will give is a modification of one given in *What Is Mathematics?* by Courant and Robbins for polygons in the Euclidean plane. This proof is neutral in that it makes no distinction between the Euclidean plane and the hyperbolic plane.

The task at hand is to identify two "components" of the complement of the polygon in $\mathbb{P}$. One is to be the interior of the polygon and the other is to be the exterior of the polygon. Intuitively, the interior should be bounded and the exterior should be unbounded. We declare a point to be interior to the polygon if a ray starting there crosses the polygon an odd number of times and exterior to the polygon if a ray starting there crosses the polygon an even number of times. We must make all this precise by showing that the parity of the number of crossings is independent of the ray chosen. We also prove that any two points in either the interior or the exterior can be connected by a polygonal path that consists entirely of points of the same kind.

Definition: Let P be a polygon, let A be a point not on P, and let $\vec{k}$ be a ray starting at A. We say that $\vec{k}$ is a *proper ray* if it does not contain any edges of P. We say that a proper ray $\vec{k}$ *crosses P through an edge E* if $\vec{k}$ intersects E and E contains points on both sides of $\vec{k}$. We say that $\vec{k}$ *crosses P at a vertex V* if V is on $\vec{k}$ and the edges of P that meet at V lie on opposite sides of $\vec{k}$. No other points of any kind are considered crossing points. The ray $\vec{k}$ is *even* if it crosses P an even number of times and *odd* if it crosses P an odd number of times.

The number of times a proper ray crosses a polygon is always finite, because the polygon has only finitely many edges and vertices. In Fig. 2.9a, Q_1 is a point where $\overrightarrow{k}$ crosses an edge, Q_2 is a point where $\overrightarrow{k}$ crosses P through a vertex, and Q_3 is not a crossing point.

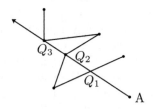

Fig. 2.9a.

Lemma: Let P be a polygon and let A be a point that is not on any line that contains an edge of P. Then every ray starting at A is proper and any two rays starting at A have the same parity.

Proof: The first assertion is easy to prove. For the second assertion, we consider rays from A in the direction of other points Q. The idea of the proof is that the parity of the crossings of $\overrightarrow{k} = \overrightarrow{AQ}$ depends continuously on Q when A is such a point, from which it follows that the parity is independent of Q. To be more precise, observe that because P is compact, there is a positive r such that no point of the closed disc $D = \{Q : |AQ| \leq r\}$ is on P. We study the behavior of the parity of $\overrightarrow{k}$ as a function of Q on the circle $C = \{Q : |AQ| = r\}$. Then continuity of the parity of $\overrightarrow{k}$ is sufficient to prove that it is constant, which we proceed to do. Let C be a crossing point of $\overrightarrow{k}$. Observe that points of $\overrightarrow{k}$ near C but on opposite sides of C are initial points of rays that have opposite parity. If C is not a vertex, there exist points C_1 and C_2 on the edge containing C such that C lies strictly between C_1 and C_2, so that Q is in the interior of the angle $\angle C_1AC_2$. Then a point Q' near enough to Q must also be in the interior of $\angle C_1AC_2$ ($|QQ'|$ less than the minimum of the distances from Q to the lines $\overline{AC_1}$ and $\overline{AC_2}$). In that case, the Crossbar Theorem says that $\overrightarrow{AQ'}$ meets the segment C_1C_2 and hence crosses P in the same segment as $\overrightarrow{k}$. If C is a vertex, similar reasoning shows that if $|QQ'|$ is small enough then $\overrightarrow{AQ'}$ either crosses P at C or crosses P through one of the edges of P that meet at C. Since $\overrightarrow{k}$ is proper, it does not contain an edge of P. Hence the only remaining possibility for a point of intersection between $\overrightarrow{k}$ and P is at a vertex V for which both edges that meet there lie on the same side of $\overrightarrow{k}$. In that case, reasoning similar to what we used in the first case shows that for Q' close enough to Q and on the same side as the two edges of P, $\overrightarrow{AQ'}$ crosses both those edges. For Q' close enough to Q and on the opposite side of $\overrightarrow{k}$ from the edges that meet at V, $\overrightarrow{AQ'}$ does not intersect P near V. This establishes the required continuity.   □

Lemma: If A is a point not on a polygon P, then any two proper rays starting at A have the same parity.

Proof: Let $\overrightarrow{k} = \overrightarrow{AQ}$ and $\overrightarrow{j} = \overrightarrow{AR}$ be proper rays starting at A. Let A' be in the ray opposite to $\overrightarrow{k}$. If $|AA'|$ is less that the distance from A to P, the parity of $\overrightarrow{A'Q}$ is the same as that of $\overrightarrow{k}$. Furthermore, a continuity argument similar to those in the proof of the preceding lemma shows that if $|AA'|$ is small enough the parity of $\overrightarrow{A'R}$ is the same as that of $\overrightarrow{j}$. In this case, one must consider points of intersection between A and R as well as those beyond R, but the basic principle is the same. Now A' is a point to which the preceding lemma applies, so the desired result for A follows.   □

Definition: If P is a polygon, its *interior* is the set of points not on P from which some ray has odd parity and its *exterior* is the set of points not on P from which some ray has even parity.

Lemma: For a triangle, this definition of interior agrees with the one in Chapter 1.

Lemma: If P is a polygon, there is a circle whose interior contains P, and every point on the circle or outside it is exterior to the polygon, while every point interior to the polygon is interior to the circle.

Lemma: Every point of the plane not on the polygon P is either interior to P or exterior to P, but not both.

Proof: If A is not on the polygon, there are only finitely many lines to avoid in order to choose rays that are proper. Hence, infinitely many of the rays that start at A are proper, each of these has a well-defined parity, and those parities are all the same.   □

Lemma: If P is a polygon then the set of interior points of P is open and the set of exterior points of P is open.

Proof: We give the proof for the interior points. The proof for the exterior points is the same. Let A be an interior point of P. Then A is not on P. Choose $r > 0$ so that the open disk $D = \{Q : |AQ| < r\}$ is disjoint from

P. There is a point A' in D that is not on any line containing an edge of P. Then every ray starting at A' is proper and cannot cross P until after it leaves D. Thus every point of D on a ray starting at A' is interior to P, and every point of D is on such a ray.   □

Lemma: Let P be a polygon and let Q_0 be a point on P. If Q_0 is an interior point of an edge E of P, then there is a positive number r such that if $|QQ_0| \le r$ and Q is on P, then Q lies on E. If Q_0 is a vertex of P, and the two edges that meet at Q_0 are E and E', then there is a positive number r such that if $|QQ_0| \le r$ and Q is on P, then Q is on E or E'. If E, E', and E'' are successive edges of P, then there is an $r > 0$ such that if $d(Q, E') \le r$ and Q is on P, then Q is in $E \cup E' \cup E''$.

Lemma: Let P be a polygon. Write U for the set of interior points of P and write W for the set of exterior points of P. If Q_0 and Q_1 are both in U then there is a polygonal path connecting Q_0 to Q_1 that lies entirely in U. A similar statement holds if both points are in W.

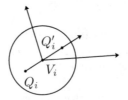

Fig. 2.9b.

Proof: Suppose the vertices of P are $V_1, \dots, V_n$. We analyze the situation near a particular vertex V_i in order to prepare for the rest of the proof. (See Fig. 2.9b.) First, choose a positive number r_i so that if $|QV_i| \le r_i$ and Q is on P, then Q is on one of the edges of P that meet at V_i. Let $D_i = \{X : |XV_i| < r_i\}$ be the open disc of radius r_i centered at V_i. We will see that the points of D_i interior to the angle $\angle V_{i-1}V_iV_{i+1}$ are either all interior to P or all exterior to P (compute subscripts modulo n), and that the points of D_i exterior to the angle are of the opposite kind. Note that there exists a point (even many points) $Q_i \in D_i$ such that the ray $\overrightarrow{Q_iV_i}$ is proper. Then by considering rays along the line $\overline{Q_iV_i}$ it can be seen that points between Q_i and V_i are on the same side of P as Q_i, while points of D_i on the ray that are on the opposite side of V_i from Q_i are on the opposite side of P from Q_i. Every point of D_i that is not on P is the initial point of at least one proper ray that meets the ray $\overrightarrow{Q_iV_i}$ before intersecting P or leaving D_i. Hence we see that the part of D_i inside the angle $\angle V_{i-1}V_iV_{i+1}$ is all interior to P or all exterior to P, and the part of D_i outside the angle is just the opposite. Let us write D_i^{int} for the part of D_i interior to P and D_i^{ext} for the part exterior to P. Choose a point Q_i' in D_i on the ray $\overrightarrow{Q_iV_i}$ and beyond V_i from Q_i. Now choose a positive number $\epsilon_i \le r_i$ small enough that every point of P whose distance from $V_{i-1}V_i$ is at most ϵ_i is in $V_{i-2}V_{i-1} \cup V_{i-1}V_i \cup V_iV_{i+1}$. Now we are ready to begin constructing a polygon interior to P that will be useful in connecting points of U by polygonal paths. Let $\epsilon = \min(\epsilon_1, \dots, \epsilon_n)$. For each i, choose a point V_i' in $Q_iQ_i' \cap D_i^{\text{int}}$ that is not on any line determined by an edge of P. Then each ray $\overrightarrow{V_i'V_{i+1}'}$ is proper. By making every $|V_jV_j'|$ small enough we can guarantee that $Q \in V_i'V_{i+1}'$ implies $d(Q, V_iV_{i+1}) < \epsilon$. Then every segment $V_i'V_{i+1}'$ lies entirely in U, so that the segments taken in order form a polygon P^{int} that is contained in U. At last we come to the construction of a polygonal path that connects Q_0 to Q_1. Let R_0' and R_1' be the points of P nearest to Q_0 and Q_1, respectively. Then we can choose points R_0 and R_1 on P close to R_0' and R_1' and on the same edges of P, respectively, so that the rays $\overrightarrow{Q_0R_0}$ and $\overrightarrow{Q_1R_1}$ are proper, and the segments Q_0R_0 and Q_1R_1 intersect P only at R_0 and R_1. Then the interior points of the segments Q_0R_0 and Q_1R_1 are contained in U. Let V_k be the vertex at one end of the edge E_0 of P containing R_0 and let V_j be a vertex at one end of the edge E_1 of P containing R_1. There is a segment connecting a point of $Q_kQ_k' \cap D_k^{\text{int}}$ to an interior point of Q_0R_0, all points of which are in U. There is a similar segment for Q_1R_1. Then segments that are parts of Q_0R_0, Q_1R_1, Q_kQ_k', Q_jQ_j', these two new segments, and segments of P^{int} can be combined to make a polygonal path in U from Q_0 to Q_1. $\square$

These lemmas combine to give the proof of the following theorem:

Theorem 2.15 (Jordan): If P is a polygon in the plane $\mathbb{P}$, then the complement of P in $\mathbb{P}$ is the disjoint union of two polygonally connected open sets, exactly one of which is bounded and is called the *interior of P*. The other open set is called the *exterior of P*.

Definition: A polygon P can be *triangulated* if the polygon and its interior can be represented as the union of a finite number of nonoverlapping triangles and their interiors. Two triangles are *nonoverlapping* if they have no interior points in common.

Theorem 2.16: Any polygon can be triangulated.

Proof: We induct on n. Any triangle is already triangulated, so suppose that $n > 3$ and that every polygon of order less than n can be triangulated. Any polygon has at least one vertex where the interior angle is less than $180°$. To see this, let A be any point of the polygon P and consider the distance function $|AP(t)|$ for t in the domain of P, $[a, b]$. This function is continuous and hence has a maximum value, say at a point t'. Set $Q = P(t')$. Then the entire polygon is contained in the closed disk of radius $r = |AQ|$ centered at A, and Q is on the boundary circle. To see that Q must be a vertex, notice first that Q is in the range of a linear part of P. That segment cannot go outside the circle, and the entire line could have at most two points on the circle. Thus Q must be an endpoint of its segment. The cases when $Q = V_n$ or $Q = V_1$ require only slightly different notation, so we suppose that $Q = V_k$, where $1 < k < n$. See Fig. 2.9c.

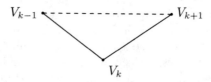

Fig. 2.9c.

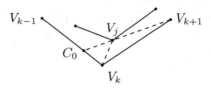

Fig. 2.9d.

If the interior points of the segment $V_{k-1}V_{k+1}$ are all interior to the polygon, we can reduce to a polygon of order $n - 1$ by using the triangle $V_{k-1}V_kV_{k+1}$ as part of a triangulation, together with a triangulation of the polygon gotten by replacing the edges involving V_k by the one segment from V_{k-1} to V_{k+1}. In the contrary case, there is a point of the polygon not on $V_{k-1}V_k$ or V_kV_{k+1} that is on $V_{k-1}V_{k+1}$ or on the same side of $V_{k-1}V_{k+1}$ as V_k, and at least one such point must be a vertex. If such vertices are all on $V_{k-1}V_{k+1}$ let V_j be one of them. Otherwise, let x be a coordinate on the line $\ell = \overline{V_kV}_{k-1}$ such that $a = x(V_k) < b = x(V_{k-1})$. There is at least one point C on ℓ such that $a < x(C) < b$ and the interior of the triangle CV_kV_{k+1} is interior to P. Choose C_0 so that $x(C_0)$ is the least upper bound of the coordinates of all such points C, as in Fig. 2.9d. Then the segment C_0V_{k+1} must intersect P at an interior point of the segment. Hence there must be a vertex of P in the interior of C_0V_{k+1}; denote that vertex by V_j. Then we can combine edges of P with the segments V_kV_j and V_jV_k to create two polygons of order less than n. If A is a point in the interior of the segment V_kV_j, then A is in U, and points near A on opposites sides of V_kV_j will both be in U, one of them in the interior of one of the new polygons and the other in the interior of the other new polygon. Since all three interiors are polygonally connected, we see that U is made up of the interiors of the two new polygons together with the interior of V_kV_j. These new polygons are of order less than n, so they can be triangulated. The segment V_kV_j will be covered by edges of triangles in each of the triangulations, so we can combine the two triangulations to get a triangulation of P.   □

2.10 Isometries; The Isometry Group

An interpretation of Axioms 5 and 6 is that all angles and all triangles obey the same laws, no matter where in the plane they are located, and no matter what their orientations. We shall see that that applies to general figures, not merely angles and triangles. We say that the plane is geometrically homogeneous and isotropic. In this section, all considerations apply equally to the Euclidean plane $\mathbb{E}^2$ and the hyperbolic plane $\mathbb{H}^2$. The symbol $\mathbb{P}$ may represent either of them. The isometry group of $\mathbb{H}^2$ is examined in greater detail beginning in Section 6 of Chapter 3, where polar coordinates are used.

General Facts About Isometries

The way we are going to make the homogeneity and isotropy explicit is by studying isometries of the plane. These are also sometimes called rigid motions of the plane. Following the definition and some observations on the definition, we investigate some particular kinds of isometries: reflections in lines, rotations about points, and translations along lines. We shall discover that every isometry can be expressed as a composition of no more than three reflections. For a given isometry, the number of reflections used must always have the same parity, and this fact will enable us to give a precise definition of the notion of orientation of $\mathbb{P}$.

Definition: A mapping σ of $\mathbb{P}$ into itself is called an *isometry* if it preserves distances. That is, if for every pair of points P, Q we have $|PQ| = |P'Q'|$, where $P' = \sigma(P)$ and $Q' = \sigma(Q)$ are the images of P and Q under the mapping. The set of all isometries of $\mathbb{P}$ will be denoted by $\mathbb{G}$.

An isometry must be one-to-one, because $P \neq Q$ implies that $|P'Q'| = |PQ| > 0$. When we know that each isometry is a composition of reflections in lines, it will follow that each isometry maps $\mathbb{P}$ onto itself. Then it is easy to verify that $\mathbb{G}$ is a group.

From the fact that an isometry σ preserves distances, it follows that σ also preserves angles. Any angle can be given by three points as $\angle PQR$, and then we can consider the triangle PQR. If $P'Q'R'$ is its image under σ, then, since $|PQ| = |P'Q'|$, etc., the sides of the first are respectively congruent to the sides of the second, and the SSS criterion (Theorem 2.7) shows that the triangles are congruent, so that $\angle P' = \angle P$, $\angle Q' = \angle Q$, $\angle R' = \angle R$.

An isometry necessarily maps lines onto lines. According to Corollary 2 of Theorem 2.2 in Section 2.3, points A, B, C are collinear, with B between A and C if and only if $|AC| = |AB| + |BC|$. Since distances are preserved, that relation holds also for $A' = \sigma(A), B' = \sigma(B), C' = \sigma(C)$, so that the primed points are collinear if the unprimed ones are.

Since geometrical properties are based on lines, distances, and angles, we see that all geometrical properties of figures are preserved by an isometry.

If $\sigma \in \mathbb{G}$, a point P is said to be *fixed* by σ provided that $\sigma(P) = P$. We are going to prove that if σ fixes three noncollinear points, then σ must be the identity transformation, that is, must fix all points. In these lemmas, we assume that $\sigma \in \mathbb{G}$.

Lemma: If σ fixes two distinct points, A and B of a line ℓ, then σ fixes every point of ℓ.

Proof: Each point C on ℓ is determined by the two numbers $|AC|$ and $|BC|$, and the hypotheses imply that $|A\sigma(C)| = |AC|$ and $|B\sigma(C)| = |BC|$.

Lemma: If there exist three noncollinear points A, B, C that are fixed by σ, then σ is the identity.

Proof: First, σ fixes all points on each of the three lines $\overline{AB}$, $\overline{AC}$, and $\overline{BC}$. Choose a point Z that is interior to the triangle. Then if X is any other point the line $\overline{XZ}$ must contain points V and W that are on $\overline{AB}$, $\overline{AC}$, or $\overline{BC}$, and hence distinct from Z. Since σ fixes both V and W, it must fix X.

Isometries in Terms of Reflections

Let ℓ be a line. There is an isometry of period two called *reflection in ℓ*, which we denote by M_ℓ and define as follows. For $A \in \ell$, $M_\ell(A) = A$. For $A \notin \ell$, let m be the line through A and perpendicular to ℓ, and let F be the intersection of ℓ and m. Let $M_\ell(A)$ be the point on m different from A whose distance from F is $|AF|$. The symmetry of the relationship makes is clear that $M_\ell(M_\ell(A)) = A$ for all A.

Lemma: For each line ℓ, M_ℓ is an isometry.

Proof: Let A, B be two points, set $A' = M_\ell(A)$, and $B' = M_\ell(B)$, and prove that $|A'B'| = |AB|$ by taking three cases: (a) $A, B \in \ell$, (b) $A \notin \ell$, $B \in \ell$, (c) $A, B \notin \ell$. The first case is clear. The second case is true by the definition of A' if B is the foot of the perpendicular from A to ℓ. Otherwise, let C be the foot of that perpendicular line and notice that the triangles ABC and $A'BC$ are congruent by the SAS criterion and the fact that $|A'C| = |AC|$. The third case is easy if the line $m = \overline{AB}$ is perpendicular to ℓ: choose a coordinate x on m that is 0 at the point where m and ℓ intersect, and use familiar facts about reflecting $\mathbb{R}$. If m is not perpendicular to ℓ, let C and D be the respective feet of perpendiculars to ℓ. First suppose that A and B are on the same side of ℓ, as in Fig. 2.10a. Then the SAS criterion shows that the triangles ACD and $A'CD$ are congruent, so that $\angle ADC$ is congruent to $\angle A'DC$. Hence, $\angle BDA = \angle B'DA'$. We also have $|AD| = |A'D|$, so the triangles ADB and $A'DB'$ are congruent, again by the SAS criterion. The case in which A and B are on opposite sides of ℓ is similar and the details are left as an exercise.

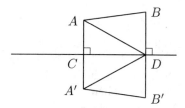

Fig. 2.10a.

The next two results are presented here because they use reflections, and will be useful later, even though they are not needed to prove that every isometry is a composition of reflections.

Lemma: Let ℓ be a line and let A be a point not on ℓ. Set $B = M_\ell(A)$. Then the half-plane of points on the same side of ℓ as A consists of the points P for which $|AP| < |BP|$.

Proof: If $P \notin \ell$, then P is either in the half-plane containing A or in the half-plane containing B. (These half-planes are different because the point C on ℓ and on the perpendicular to ℓ through A is in the segment AB.) By symmetry, it will suffice to prove that $|AP| < |BP|$ in the first case. In that case, the segment PB contains a point Q of ℓ. Then $|BP| = |BQ| + |QP|$, $|BQ| = |AQ|$, and the points A, P, and Q are not collinear, so $|AP| < |AQ| + |QP|$.

Corollary: Half-planes are open sets.

Lemma: If σ is not the identity, but fixes every point on some line ℓ, then $\sigma = M_\ell$.

Proof: There must be a point A for which $\sigma(A) \neq A$. Let m be the line through A perpendicular to ℓ, meeting ℓ in the point F. Since $\sigma(\ell) = \ell$, and σ must preserve angles, $\sigma(m) \perp \ell$. Since $\sigma(F) = F \in m$, $\sigma(m) = m$. Thus $\sigma(A)$ is a point of m that is distinct from A but the same distance from F as A. Thus $\sigma(A) = M_\ell(A)$, so $M_\ell \circ \sigma$ fixes A and all points on ℓ. Hence $M_\ell \circ \sigma$ is the identity, which implies the desired conclusion.

Lemma: If ℓ is a line, $\sigma \circ M_\ell \circ \sigma^{-1} = M_{\sigma(\ell)}$.

Proof: Just calculate to see what each isometry does to points on and off $\sigma(\ell)$.

Lemma: Suppose that σ fixes a point P, but does not fix all the points on any line. If A is any point different from P and $\ell = \overline{AP}$, then there is a unique line m through P such that $\sigma = M_m \circ M_\ell$.

Remark: An isometry of this type is called a *rotation about P*.

Proof: If $\sigma \circ M_\ell$ is a reflection, its line is determined, so the existence is the main part of the proof. Let A be a point distinct from P and set $B = \sigma(A)$. Then $B \neq A$. Let m be the line bisecting $\angle APB$, and let C be the intersection of m with $\overline{AB}$. Then $|PA| = |PB|$ because σ fixes P and takes A to B, so the SAS criterion shows that the triangles PAC and PBC are congruent. Hence $|AC| = |BC|$ and $\angle PCA = \angle PCB$. Since the two latter angles are supplementary, they must be right angles. This proves that $M_m(A) = B$, and hence that $M_m \circ \sigma$ fixes both P and A. By hypothesis, σ is not a reflection, so $M_m \circ \sigma$ is not the identity, but the fact that it fixes P and A then implies that $M_m \circ \sigma$ is $M_{\overline{PA}}$, which implies the desired conclusion.

Theorem 2.17: Every isometry σ that is not the identity is a product of one, two, or three reflections.

Proof: By the lemmas, we may assume that σ has no fixed points. Choose a point A and define $B = \sigma(A)$. Let ℓ be the perpendicular bisector of the segment AB. Then $M_\ell \circ \sigma$ fixes A, but is not the identity, because then σ would be a reflection and hence have fixed points. If $M_\ell \circ \sigma$ fixes a point other than A, then it must be a reflection, so σ is a product of two reflections. If $M_\ell \circ \sigma$ fixes only A, then it is a rotation and σ is a product of three reflections.

The method used to prove Theorem 2.17 can be used to establish two forms of the homogeneity mentioned at the beginning of this section:

Lemma: If AB and DE are congruent segments, there are two reflections whose composition takes A to D and B to E.

Theorem 2.18: If ABC and DEF are congruent triangles, then there is a unique isometry σ such that $\sigma(A) = D$, $\sigma(B) = E$, and $\sigma(C) = F$.

We know that two reflections in intersecting lines combine to have a fixed point. What about lines that don't intersect?

Lemma: Let ℓ and m be nonintersecting lines and set $\sigma = M_m \circ M_\ell$. Then σ has no fixed points.

Proof: If P is a point on the opposite side of ℓ from m, then

$$d(\sigma(P), m) = d(M_\ell(P), m) < d(P, m),$$

so $\sigma(P) \neq P$. Similar reasoning will provide the proof if P is on the same side of ℓ as m, and points on ℓ are clearly moved.

Lemma: Let ℓ and m be distinct lines that are perpendicular to a third line, n. Set $\sigma = M_m \circ M_\ell$ and let x be a coordinate function on n. Suppose that ℓ intersects n at a point A and that m intersects n at a point B. Then for $P \in n$,

$$x(\sigma(P)) = x(P) + 2(x(B) - x(A)).$$

Remark: Such an isometry is called a *translation along n* (by $2(x(B) - x(A))$).

Proof: It suffices to verify the formula for a single point of n, because of what is known about isometries of $\mathbb{R}$, and it is clear for the point A.

Lemma: Let n be a line, HP a half-plane bordered by n, and σ an isometry that takes n to n, HP to HP, and has no fixed points on n. Let A be any point of n and let ℓ be the line perpendicular to n at A. Then there is a unique line m perpendicular to n such that $\sigma = M_m \circ M_\ell$ so that σ is a translation along n.

Orientation

A reflection has a line of fixed points, while a product of two reflections has a single fixed point if the two lines intersect but are not identical, and no fixed points at all if the two lines do not intersect. Hence, if σ is a product of two reflections it is not itself a reflection. In order to prove that a composition of an even number of reflections is never equal to a composition of an odd number of reflections, we could develop information about the compositions of translations and rotations enabling us to reduce the number of reflections involved. Instead, we are going to introduce the notion of an orientation of $\mathbb{P}$ and prove that reflections reverse orientation. There are clear intuitive notions of orientation that we need to make rigorous. We need to know that "right-handed" and "left-handed" orientations are different, that is, that there are exactly two possible choices of orientation. We can describe these intuitively in terms of handedness or in terms of clockwise or counter-clockwise rotations, but mathematically one must simply choose one of two possible orientations. Pictures on a sheet of paper fit the natural intuition about orientation, and Figs. 2.10b and 2.10c illustrate the concept graphically.

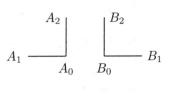

Fig. 2.10b.

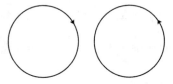

Fig. 2.10c.

In order to make the notion of orientation rigorous, begin by introducing coordinates in $\mathbb{P}$: choose a line ℓ and a coordinate function x on ℓ, and choose HP to be one of the two half-planes determined by ℓ. Define $\varphi = (\varphi_1, \varphi_2)$ from $\mathbb{P}$ to $\mathbb{R}^2$ as follows: if A is a point of HP, let F be the foot of the perpendicular from A to ℓ and let $\varphi(A) = (x(F), |AF|)$. If A is in the opposite half-plane from HP, replace $|AF|$ by $-|AF|$. If $A \in \ell$, let $\varphi(A) = (x(A), 0)$. It is easy to see that φ is one-to-one and onto. Theorem 2.13 can be used to prove that the first component of φ is continuous, and the continuity of the second component follows from Corollary 3 of Theorem 2.2. We can use this function φ to distinguish "right-handed" and "left-handed" orientations. A different choice of oriented line or half-plane could switch the two categories. Our current need is just to provide some way of distinguishing two orientations. On general manifolds, it is usually necessary to compare orientations of two coordinate systems by taking the determinant of a matrix of partial derivatives of one set of coordinate functions with respect to the other. Because we are in such a special situation, the derivatives are not necessary, but we still use the standard test of positivity of a determinant. We are going to test everything against the one coordinate system introduced above.

Let us use the term *frame* for an ordered triple of points, (A_0, A_1, A_2), with the property that $|A_0A_1| = |A_0A_2| = 1$, and such that the angle $\angle A_1 A_0 A_2$ is a right angle. We can make a metric (distance function) on the set of all frames by saying the distance between two frames (A_0, A_1, A_2) and (B_0, B_1, B_2) is $|A_0B_0| + |A_1B_1| + |A_2B_2|$. It is straightforward to check that this function is symmetric, nonzero on distinct frames, and satisfies the triangle inequality.

If we begin to choose a frame by choosing two points A_0 and A_1 that are a distance 1 apart, the third point to complete a frame must be on the line perpendicular to $\overline{A_0A_1}$ at a distance of 1 from A_0. Thus there are exactly two ways to form a frame from such a pair. Our objective is to provide a clear criterion for distinguishing the two choices. In the process, it will be useful to know a fact about the situation in which $\varphi_2(A_1) = \varphi_2(A_2)$.

Lemma: Let m be a line and let A and B be points in the same half-plane of m and at the same distance from m. Suppose that $|AB| < 1$. If P is a point such that $|AP| = |BP| = 1$ (there are two such points, by Theorem 2.11), then the foot of the perpendicular from P to m lies between the feet of the perpendiculars from A and B.

Proof: Let E be the foot of the perpendicular from A to m and let F be the foot of the perpendicular from B to m. Then the triangles AEF and BFE are congruent by the SAS criterion. Using the SAS criterion again, we see that the triangles ABF and BAE are congruent, so that the angles $\angle ABF$ and $\angle BAE$ have the same measure. Let C be the midpoint of the segment AB and let D be the midpoint of the segment EF. Then the triangles CAE and CBF are congruent, so that ECF is isosceles. Hence CD is perpendicular to m. Now the point P must lie on the line $\overline{CD}$, so that D is the foot of its perpendicular to m.

Intuitively speaking, if rotating from A_1 to A_2 is counter-clockwise, we call the frame "right-handed" and the other case is "left-handed." To get something precise, first define

$$\Delta_1(A_0, A_1, A_2) = \varphi(A_1) - \varphi(A_0)$$

and

$$\Delta_2(A_0, A_1, A_2) = \varphi(A_2) - \varphi(A_0).$$

These are both pairs of real numbers, so we can form a 2-by-2 matrix using $\Delta_1(A_0, A_1, A_2)$ as the first row and $\Delta_2(A_0, A_1, A_2)$ as the second row. Let $\Delta(A_0, A_1, A_2)$ be the determinant of the resulting matrix. Observe that Δ is continuous: it is a polynomial in the coordinates of the points that are in the frame.

Theorem 2.19: $\Delta(A_0, A_1, A_2)$ is never 0.

Remark: We call the frame (A_0, A_1, A_2) *positive* (relative to the coordinates determined by a coordinate on ℓ and a choice of half-plane) provided $\Delta(A_0, A_1, A_2) > 0$, and *negative* in the opposite case.

Proof: For the determinant to be 0, one row must be a multiple of the other. To see that the latter is impossible, use a consequence of the condition $|A_0 A_1| = 1$. By the preceding lemma, we see that for each point $y \in [\varphi_2(A_0) - 1, \varphi_2(A_0) + 1]$ there is exactly one point A such that $|A_0 A| = 1$, $\varphi_2(A) = y$, and $\varphi_1(A) > \varphi_1(A_0)$. If there were two such points, we could take them to be the A and B of the lemma, and A_0 to be the P. In other words, $\varphi_1(A)$ is determined by $\varphi_2(A)$, if $\varphi_1(A) > 0$. A similar statement is true for points A such that $\varphi_1(A) < 0$. Hence the only possible multiples for the second row to be of the first are 1 and -1. But if $\varphi_2(A_2) = \varphi_2(A_1)$, then $\varphi_1(A_2) = -\varphi_1(A_1)$ (and there are only two such configurations).

Continuous Curves of Isometries

We already have the notion of a continuous curve in the plane $\mathbb{P}$, and want to extend the notion to curves in the isometry group $\mathbb{G}$. (There could even be continuous functions from other spaces to $\mathbb{G}$.) If $[a, b]$ is an interval in $\mathbb{R}$, and we have an isometry σ_t for each $t \in [a, b]$, we say this is a *continuous curve in* $\mathbb{G}$ if the function taking t to $\sigma_t(P)$ is a continuous curve in $\mathbb{P}$ for every point P.

Theorem 2.20: If σ_t is a continuous curve in $\mathbb{G}$, defined on $[a, b]$, and (A_0, A_1, A_2) is a frame, then $(\sigma_t(A_0), \sigma_t(A_1), \sigma_t(A_2))$ has the same orientation for all values of t.

Proof: This is immediate from the continuity of the determinant function used to define the orientation.

Our next goal is to show that continuous curves in $\mathbb{G}$ can be constructed to achieve certain motions.

Lemma: Let A and B be distinct points, set $\ell = \overline{AB}$ and let HP be one of the half-planes determined by ℓ. Let $\mathcal{R}_A$ be the set of rays in HP starting at A and let $\mathcal{R}_B$ be the analogous set for B. Consider the one-one correspondence between subsets of $\mathcal{R}_A$ and $\mathcal{R}_B$ that associates two rays iff they meet at right angles. For $\vec{j} \in \mathcal{R}_A$, let α be the angle between $\overrightarrow{AB}$ and $\vec{j}$, and for the ray $\vec{k} \in \mathcal{R}_B$ perpendicular to $\vec{j}$, let β be the angle between $\overrightarrow{BA}$ and $\vec{k}$. Then β is a strictly decreasing continuous function of α.

Proof: Let f be the function defined in the theorem: for $0 < \alpha < \pi/2$, the ray starting at A with angle α meets exactly one ray starting at B at a right angle, and we write $f(\alpha)$ for the angle at B. (See Fig. 2.10d.) Take α_0 in that range, set $\beta_0 = f(\alpha_0)$, denote the corresponding rays by $\vec{j}_0$ and $\vec{k}_0$, and call their point of intersection Z. Let $\epsilon > 0$, and choose β' so that $\beta_0 - \epsilon < \beta' < \beta_0$. Let $\vec{k}$ be the ray at B with angle β', which intersects $\vec{j}_0$ at a point X such that the angle $\angle AXB$ is greater than a right angle. (Consider first the triangle BZX, which has one right angle.) Then there is an angle $\alpha_1 > \alpha_0$ such that if $\vec{j}$ is a ray at A of angle between α_0 and α_1 then $\vec{j}$ also meets $\vec{k}$ at an angle greater than $90°$. It follows that if $\alpha_0 < \alpha < \alpha_1$ then $\beta_0 - \epsilon < f(\alpha) < \beta_0$. This proves that f is continuous from the right. To see that f is continuous from the left, a similar argument can be given.

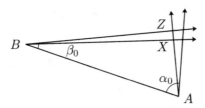

Fig. 2.10d.

Lemma: Take the situation and notation of the preceding lemma, and denote the point of intersection of corresponding rays by Q. Then $|BQ|$ is an increasing continuous function of α, and Q is a continuous function of α.

Proof: In the previous proof, the point Z was the Q for α_0. The Q for an α between α_0 and α_1 is in the quadrilateral bounded by the four rays named in the proof. The continuity of a point of intersection of a ray with a fixed line as a function of the angle the ray makes with another fixed line allows us to show that $|ZQ|$ can be made as small as desired by making α_1 close to α_0. A similar argument can be used for left continuity.

Lemma: If A, B, P are three points, set $\ell = \overline{AB}$ and define $f(A, B, P) =$

$M_\ell P$. The function f is a continuous function of the three points.

Proof: It suffices to prove that f is continuous as a function of each individual variable separately. The continuous dependence on P follows from the fact that M_ℓ is an isometry. To discuss the dependence of f on B for a fixed A and P, apply the preceding lemma to the pair A and P in place of A and B. If B is changed so that the angle α changes, the point Q in that lemma is the foot of the perpendicular from P to $\overline{AB}$ and since that varies continuously, so does the image of P.

Lemma: Let A and B be any two points on a line ℓ. Then there is a continuous curve, σ_t for $t \in [0, |AB|]$, consisting of translations along ℓ, such that σ_0 is the identity and $\sigma_{|AB|}(A) = B$.

Proof: Choose a coordinate function x on ℓ such that $x(A) = 0$ and $x(B) > 0$. For each t, let m_t be the line perpendicular to ℓ at the point whose coordinate is $t/2$. Set $\sigma_t = M_{m_t} \circ M_{m_0}$, and verify that this σ_t does what is needed.

Lemma: Let P be any point and let A and B be any two points at the same distance from P. Then there is a continuous curve of rotations about P that starts at the identity and whose endpoint carries A to B.

Theorem 2.21: Let AB and CD be any two congruent segments. Then there is a continuous curve of isometries that starts at the identity and whose endpoint carries A to C and B to D.

Theorem 2.22: Let (A_0, A_1, A_2) and (B_0, B_1, B_2) be two frames that have the same orientation. Then there is a continuous curve in $\mathbb{G}$ starting at the identity, whose endpoint carries (A_0, A_1, A_2) to (B_0, B_1, B_2).

Theorem 2.23: If ℓ is a line and (A_0, A_1, A_2) is any frame, then (A_0, A_1, A_2) and its image under M_ℓ have opposite orientations.

Proof: We can reduce to proving that M_ℓ reverses orientation for frames whose first two points lie on ℓ. Then M_ℓ switches the two possible third components and thus reverses orientation, as claimed.

Corollary: A product of an odd number of reflections reverses orientation and a product of an even number of reflections preserves orientation. Thus a product of an even number of reflections is never a product of an odd number of reflections.

Corollary: Translations and rotations preserve orientation.

Chapter 3

Qualitative Description of the Hyperbolic Plane

The hyperbolic plane is qualitatively different from the Euclidean plane in a number of ways. Among these are the fact that the sum of the angles of a triangle is strictly less than π (radians). The difference between π and the sum is called the *defect* of the triangle and is proportional to the area of the triangle, so the areas of triangles are bounded above. It is true in the hyperbolic plane that the length of a circular arc and the area of a circular sector are both proportional to the angle of the arc. What is different is the way the proportionality depends on the radius of the circle. Another difference is in the structure of the group of isometries, which are mappings of the plane that preserve distance. Since tilings and lattices are tied so closely to the isometries of the plane, they are also very different from the Euclidean case. All of the distinctive properties can be traced to the hyperbolic parallel axiom, so we begin with the theory of parallelism, presented in roughly the form in which Gauss derived it.

3.1 The Angle of Parallelism and Asymptotic Pencils

We now assume the validity of the hyperbolic parallel axiom, Axiom 7b; hence, one can find a point A and line m for which at least two lines ℓ and ℓ' through A do not intersect m. Let A, m be such, and let B be the foot of the perpendicular dropped from A to m. Of the four rays from A contained in those two lines, it is clear, by consideration of supplementary angles, that at least one of those rays, say ray $\overrightarrow{k}$, makes an acute angle θ_0 with the ray $\overrightarrow{AB}$, as indicated in Fig. 3.1a, in which we have also indicated by $\overrightarrow{m}$ the ray from B in the line m that is on the same side of the line AB as the ray $\overrightarrow{k}$. If a ray from A at an angle θ_2 intersects m, then any ray at any angle $\theta_1 < \theta_2$ also intersects $\overrightarrow{m}$, by the Crossbar Theorem, as in Fig. 3.1b. Therefore, if α is the least upper bound of the angles θ such that

a ray at angle θ intersects $\overrightarrow{m}$, then all rays at angles less than α intersect $\overrightarrow{m}$ and all rays at greater angles do not intersect $\overrightarrow{m}$. We call α the *angle of parallelism* (for the point A and the line m); α is $\leq \theta_0$, hence less than $\pi/2$.

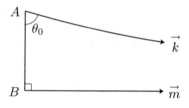

Fig. 3.1a.

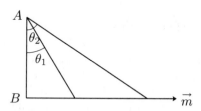

Fig. 3.1b.

It is easy to see that the ray *at* the angle α also does not intersect $\overrightarrow{m}$, for it it did, say at a point C on $\overrightarrow{m}$, and if D were a point farther away on $\overrightarrow{m}$, then the ray $\overrightarrow{AD}$ would be at an angle greater than α and would intersect $\overrightarrow{m}$ (namely, at D), and that would contradict the statement that α is an *upper bound* of the angles of rays that intersect $\overrightarrow{m}$. The ray at angle α is *asymptotic* to $\overrightarrow{m}$.

We now consider any other segment $A'B'$ in the plane, of the same length as AB, and a line m' through B' perpendicular to $A'B'$. Let $\overrightarrow{m'}$ be one of the two rays (i.e., in either direction) from B' along m' as in Fig. 3.1c. If C is any point on $\overrightarrow{m}$ and C' is the point on $\overrightarrow{m'}$ at the same distance from B' as C is from B, then, by the SAS axiom, the triangles ABC and $A'B'C'$ are congruent, hence $\angle BAC = \angle B'A'C'$. Therefore, a ray from A at an angle θ intersects $\overrightarrow{m}$ if and only if the ray from A' at the same angle θ intersects $\overrightarrow{m'}$. Therefore, the entire discussion above applies also to the figure with primed symbols; hence the angle of parallelism for A' and m' is the same as that for A and m.

We have proved the following result:

Theorem 3.1: The angle α of parallelism of a point A with respect to a line m depends only on the distance $y = |AB|$ of the point A from m. (That is, in this respect, at least, the hyperbolic plane is homogeneous and isotropic throughout.)

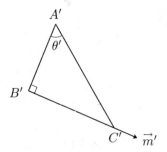

Fig. 3.1c.

We denote by $\Pi(y)$ the angle α as a function of y.

As a special case of the above argument, if AB is the same segment as $A'B'$, while $\overrightarrow{m}$ and $\overrightarrow{m}'$ are opposite rays in the line m, we see that there are two rays through A, one on each side, which are asymptotic to m, both making angle α with respect to $\overrightarrow{AB}$, as in Fig. 3.1d.

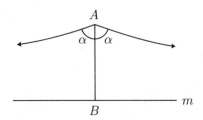

Fig. 3.1d.

If ℓ, m are directed lines that contain rays $\overrightarrow{\ell}, \overrightarrow{m}$ (it being understood that in each case, the direction of the ray agrees with the direction of the line), we say that ℓ is *asymptotic* to m if $\overrightarrow{\ell}$ is asymptotic to $\overrightarrow{m}$, and we wish to show that being asymptotic is a property of the directed lines, and is independent of the choice of the point A on ℓ.

Theorem 3.2: Let ℓ and m be directed lines and A, A' points on ℓ. Then ℓ is asymptotic to m through A if and only if it is also asymptotic to m through A'.

Proof: Suppose that A' is beyond A (in the direction of increasing coordinate on ℓ) as in Figs. 3.1e and 3.1f. (Otherwise, interchange A and A'.) Then, assume first that ℓ is asymptotic to m through A. To show that it

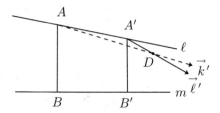

Fig. 3.1e.

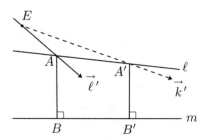

Fig. 3.1f.

is also asymptotic to m through A', we note that if there were a ray $\overrightarrow{\ell}'$ below ℓ on the right from A' that does not intersect m, we could take a point D on it, and then a ray $\overrightarrow{k}'$ from A through D (shown dashed) would not intersect m (because points of $\overrightarrow{k}'$ beyond D are on opposite sides of $\overrightarrow{\ell}'$ from A, hence from B, hence from all of m). That would give a contradiction, because the ray $\overrightarrow{k}'$, since it lies below the asymptotic line ℓ, would necessarily intersect m. Therefore, there is no ray from A' below ℓ that fails to intersect m, while ℓ itself does not intersect m. Therefore, ℓ is asymptotic to m through A' (as well as through A). Now assume that ℓ is asymptotic to m through A'. The argument is similar; if there were a ray $\overrightarrow{\ell}'$ through A below ℓ that does not intersect m, as in Fig. 3.1f, we could extend it backward and choose a point E on it to the left of A. Then the ray from E through A' would not intersect m, and that would contradict the assumption that ℓ is asymptotic to m through A'. □

We now wish to show that the relation of being asymptotic is symmetric. If ℓ is asymptotic to m, then m is asymptotic to ℓ. We recall from Theorem 2.2 that if A is any point and m any line, and x is a coordinate on m, then the distance $|AP|$ from A to a point P on m with coordinate

x is a continuous function of x. That function has a single minimum and increases with increasing $|x|$ as P moves to infinity in either direction.

In the proof of the next theorem, we make use of Corollary 3 to Theorem 2.2, according to which the distance $d(C, \ell)$ of a point C from a line ℓ is a continuous function of the position of C in the plane, so that, in particular, if C varies on a line k, then $d(C, \ell)$ is a continuous function of a coordinate of C on the line k.

Theorem 3.3: Let ℓ and m be directed lines. Then ℓ is asymptotic to m if and only if m is asymptotic to ℓ.

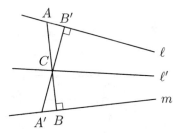

Fig. 3.1g.

Proof: Drop a perpendicular from a given point A on ℓ to m, as in Fig. 3.1g. Let C be an intermediate point on AB (to be chosen in a moment), and let $A'B'$ be a segment passing through C and perpendicular to ℓ, as shown. As the point C moves downward from A to B, the distance $|CB|$ decreases to zero, while the distance $|CB'|$ is, as noted above, a continuous function of $|CB|$, which is zero when C coincides with A and is positive otherwise. Therefore, by continuity, the distances $|CB|$ and $|CB'|$ are equal for some choice of C. The ASA criterion for congruence implies that the triangles $AB'C$ and $A'BC$ are congruent, so that $|A'B'| = |AB|$ and $\angle CAB' = \angle CA'B$ since $\angle CAB'$ is the angle of parallelism for the distances $|AB|$, then angle $\angle CA'B$ is the angle of parallelism for the (same) distance $|A'B'|$ so it follows that m is asymptotic to ℓ. Hence $d(A', \ell) \to 0$ as $A' \to \infty$ on m.   □

(We note in passing that it is essential to have directions specified on the lines. Lines that are asymptotic in one direction are not asymptotic in the opposite direction. In Fig. 3.1d, the two rays from A asymptotic to m in the two directions do not lie in a single line, because the angle of parallelism α is less than $\pi/2$. We shall see that asymptotic lines get closer together in the specified direction and farther apart in the other.)

Last, we show that the asymptotic relation is transitive, as stated in the theorem below. We need first a lemma.

Lemma 3.1: If distinct directed lines k and m are both asymptotic to directed line ℓ, then one of the three lines lies between the other two. That is, the other two lie in opposite half-planes bounded by the first line.

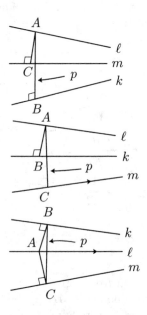

Fig. 3.1h.

Proof: We note first that the lines k and m don't intersect, for if they did, we could drop a perpendicular from the intersection to ℓ. Then, k and m would both make the same angle with respect to that perpendicular (and on the same side of that perpendicular), namely, the angle of parallelism, hence k and m would coincide, but we have assumed them distinct. We now see that none of the three lines intersects either of the other two: k and ℓ don't intersect, because they are asymptotic, and the same holds for the lines m and ℓ. We show next that we can find a line p that intersects all three given lines. From an arbitrary point A of ℓ, we drop perpendiculars AB and AC to k and m. If points B and C are on opposite sides of line ℓ, as in the bottom drawing of Fig. 3.1h, then segment BC intersects line ℓ, by Axiom 4, and BC is the desired line p. If B and C are on the same side of line ℓ, as in either of the top drawings, we proceed as follows: Let $\overrightarrow{\ell}$ be a ray from A in the direction of the line ℓ. The angles $\angle AB$, $\overrightarrow{\ell}$ and $\angle AC$,

$\overset{\rightarrow}{\ell}$ are on the same side of ℓ. If those angles are equal, then points A, B, C, are collinear, and their line is the desired line p. If ray $\overset{\longrightarrow}{A\,B}$ lies in the interior of the angle $\angle AC$, $\overset{\rightarrow}{\ell}$ as in the top drawing of Fig. 3.1h, then that ray intersects m, by definition of asymptotic lines, because ℓ is asymptotic to m through A; hence the line of AB is the desired line. In the other case, where the ray $\overset{\longrightarrow}{AC}$ lies in the interior of the angle $\angle AB$, $\overset{\rightarrow}{\ell}$, as in the middle drawing of Fig. 3.1h, is similar. Now let K, L, M be the intersections of k, ℓ, m with the line p. One of those points lies between the other two on the line p. Suppose, for example, that L lies between K and M. Then, K and M are on opposite sides of the line ℓ, and, from the obvious fact that if two lines do not intersect, any two points of the first are on the *same* side of the second, because the segment joining them does not intersect the second line, we see that the entire lines k and m are on opposite sides of ℓ. The other cases are similar, and that completes the proof of the lemma. □

(We note in passing that the lemma would be false if we had not assumed that k and m are asymptotic to ℓ. In the hyperbolic plane, it is possible for three nonintersecting lines to be such that none of them is between the other two, as indicated schematically in Fig. 3.1i.)

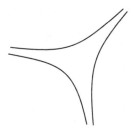

Fig. 3.1i.

Theorem 3.4: Let k, ℓ, m be directed lines. If k and m are both asymptotic to ℓ, then they are asymptotic to each other.

Proof: We have to consider two cases. First, suppose that ℓ lies between k and m, as in Fig. 3.1j. From a point A of k, any ray $\overset{\rightarrow}{k'}$ below k intersects ℓ at a point, say B. Past B, $\overset{\rightarrow}{k'}$ lies below ℓ, hence eventually crosses m, because ℓ is asymptotic to m. That is, any ray $\overset{\rightarrow}{k'}$ from A that lies between k and m intersects m. It follows that k is asymptotic to m, as was to be shown. Now suppose that one of k, m is between the other two, say k is between ℓ and m, as in Fig. 3.1k. Then, from a point of m any ray $\overset{\rightarrow}{m'}$ above

m crosses ℓ, because m is asymptotic to ℓ; hence it crosses k, because m and ℓ are on opposite sides of k. Therefore m is asymptotic to k.   □

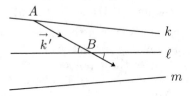

Fig. 3.1j.

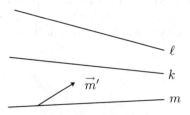

Fig. 3.1k.

In order to make the asymptotic relation reflexive, we define a directed line as being *asymptotic to itself*. Then the relation is symmetric, reflexive, and transitive.

Definition 3.1: The set of all directed lines that are asymptotic to a given directed line is called an *asymptotic pencil of lines*. They are all asymptotic to each other.

3.2 Angular Defects of Triangles and Other Polygons

According to Theorem 2.12, the sum of the angles of a triangle is less than or equal to π. We shall see later in this section that it is always less in hyperbolic geometry, but in any case we define the *angular defect* (or often simply the *defect*) of the triangle as π minus the sum of its angles. Suppose a triangle is divided into two smaller triangles by a line from one vertex to a point on the opposite side, as in Fig. 3.2a. Since $\gamma + \delta = \pi$, the sum of the defects of the smaller triangles is $2\pi - (\alpha + \beta + \gamma + \delta + \varepsilon + \zeta) = \pi - \alpha - (\beta + \varepsilon) - \zeta$, and that is the defect of the large triangle. We have proved the following theorem.

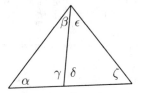

Fig. 3.2a.

Theorem 3.5: When a triangle is decomposed as the union of two smaller triangles, as in Fig. 3.2a, the angular defect of the large triangle is equal to the sum of the defects of the smaller triangles.

We mention a grand generalization of this, according to which if a large polygon is decomposed as the union of a finite number of smaller polygons, then the defect of the large polygon is the sum of the defects of the smaller polygons. We shall not prove this in all generality, but consider some special cases. First, of course, we have to define the angular defect of a polygon.

We consider first a convex n-sided polygon or n-gon. From an interior point Z we draw lines connecting Z to the vertices of the n-gon. Since the n-gon is convex, these connecting lines are all interior and they divide the n-gon into n triangles, as indicated schematically in Fig. 3.2b.

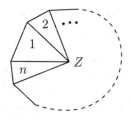

Fig. 3.2b.

Let the angles of the ith triangle be $\alpha_i, \beta_i, \gamma_i$, in clockwise order around the triangle such that γ_i is the vertex at Z. An (interior) angle of the n-gon is of the form $\beta_i + \alpha_{i+1}$, except for the last, which is $\beta_n + \alpha_1$. Hence the sum of the angles of the n-gon is $\sum_{i=1}^{n}(\alpha_i + \beta_i)$. Since the sum of the angles γ_i at Z is 2π, we can write

$$\text{sum of the angles of the } n\text{-gon} = \sum(\alpha_i + \beta_i + \gamma_i) - 2\pi.$$

If δ_i denotes the defect of the ith triangle, we have $\alpha_i + \beta_i + \gamma_i = \pi - \delta_i$. We therefore make the following definition.

Definition 3.2: The *defect* of the n-gon is $(n-2)\pi -$ the sum of its angles.

Theorem 3.6: When a convex n-gon is the union of n triangles as in Fig. 3.2b, its angular defect is equal to the sum of the defects of the triangles.

The theorem is obviously still true if we include degenerate vertices, by which we mean points in otherwise straight sides, as at A in Fig. 3.2c. The angle of the polygon at such a vertex is defined to be π.

Fig. 3.2c.

We now suppose that we have one convex polygon inside another. Let Z be a point in the interior of the inner polygon and draw lines from Z to all the vertices of the two polygons (thus generally creating some degenerate vertices where these lines meet the sides of the polygons). Let one of the triangles into which the outside polygon is decomposed be AEZ as in Fig. 3.2d, where BDZ is a corresponding triangle of the inside polygon, as in Fig. 3.2e.

We draw the line BE. The defect of EBZ is the sum of the defects of the triangles labeled 1 and 2, and then the defect of AEZ is the sum of that defect and the defect of the triangle labeled 3. Consequently, the defect of BDZ is less than or equal to the defect of AEZ. By adding together contributions from all the triangles, we have the following result.

Theorem 3.7: If one convex polygon lies inside another convex polygon, then the defect of the first is less than or equal to the defect of the second.

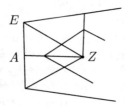

Fig. 3.2d.

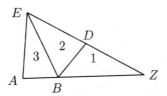

Fig. 3.2e.

For a nonconvex polygon with a reentrant vertex at a point Q, as in Fig. 3.2f, we define the (*interior*) *angle* at Q to be equal to 2π minus the exterior angle α. Then, the above theorem continues to hold when the words "convex" are omitted. The proof is similar to the one above, but a number of special cases have to be considered. The reader will surely be able to give proofs for many of those special cases. We shall see later that in the hyperbolic plane angular defects are never zero, hence "less than or equal" in the above can always be replaced by "less than."

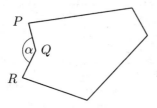

Fig. 3.2f.

Theorem 3.8: There exists in the hyperbolic plane a triangle with nonzero (hence positive) angular defect, that is, with angle sum less than π.

Proof: Recall that the angle of parallelism $\alpha = \Pi(y)$ is less than $\pi/2$ for certain values of y. (It is less than $\pi/2$ for all y, but that has not been proved yet.) Let AB be perpendicular to a line ℓ and of length y, as in Fig. 3.2g. Let ε be an angle less than α and less than $\pi/2 - \alpha$, and let C be such that AC makes angle $\alpha - \varepsilon$ with AB, as shown. Let D be a point sufficiently far to the right on ℓ that $|AC| = |CD|$. Then the triangle ACD is isosceles. The angle CAD is less than ε, hence the angle CDA is also less than ε. The angle BAD is less than α, the angle of parallelism; hence the angle sum of the large triangle ABD is less than $\pi/2 + \alpha + \varepsilon$. Thus the defect of ABD is greater than the positive number $\pi/2 - \alpha - \varepsilon$, and in fact can be made as close as we wish to the positive quantity $\pi/2 - \alpha$, by taking ε small.　　　　□

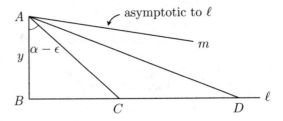

Fig. 3.2g.

Theorem 3.9: In the hyperbolic plane, *every* triangle has positive defect.

Proof: We shall show that if there were a triangle with zero defect, then all triangles would have zero defect, and that would contradict the preceding

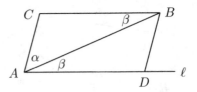

Fig. 3.2h.

theorem. Assume that triangle ABC in Fig. 3.2h, with angles α and β has zero defect. From point A draw a ray $\overrightarrow{\ell}$ at angle β with $\overrightarrow{AB}$ and so that $\overrightarrow{\ell}$ and AC are on opposite sides of AB, and on that ray $\overrightarrow{\ell}$ choose D so that $|AD| = |CB|$. Then, by the SAS criterion, the two triangles shown in Fig. 3.2h are congruent, hence both have zero defect, and the quadrilateral $ABDC$ has zero defect, equal opposite angles, and equal opposite sides, that is, it would be a parallelogram. By placing copies of that parallelogram end to end and side by side, we could construct an arbitrarily large paralellogram with zero defect. Then, given any triangle whatever, we could choose the paralellogram large enough to contain it; hence, by Theorem 3.7, the triangle would have zero defect. That is, all triangles would have zero defect, and that would contradict Theorem 3.8, according to which there exists at least one triangle with positive defect.   □

In Section 3.9 below it will be seen that the angular defect of a triangle is proportional to its area. Hence, very small triangles have very small defects, as already seems evident from Theorem 3.5 above. Here we prove a partial result.

Theorem 3.10 (Thin Triangles): If the length of the shortest side of a triangle tends to zero, while the lengths of the other sides are kept below a fixed bound, then the angular defect tends to zero.

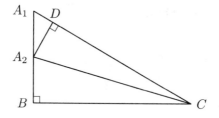

Fig. 3.2i.

Proof: In the right triangle $A_1 BC$ shown in Fig. 3.2i, we bisect the angle at C, we call A_2 the intersection of the bisector and side $A_1 B$, and we drop a perpendicular from A_2 to $A_1 C$ with foot at D. Then the triangles $A_2 DC$ and $A_2 BC$ are congruent by the AAS criterion. The defect of $A_2 DC$ is less than that of $A_2 A_1 C$; hence the defect of $A_2 BC$ is less than half that of $A_1 BC$. We then bisect the angle at C of the triangle $A_2 BC$ similarly, and so on. The defect of $A_n BC$ tends to zero, as $n \to \infty$, and any triangle that can fit inside $A_n BC$ has a still smaller defect, and the theorem follows.   □

3.3 Application to the Angle of Parallelism

Theorem 3.11: The function $\Pi(y)$ is continuous and decreasing, for $0 < y < \infty$. Its limiting values are $\pi/2$ as $y \to 0$ and 0 as $y \to \infty$.

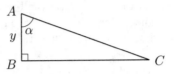

Fig. 3.3a.

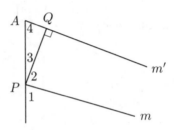

Fig. 3.3b.

Proof: According to Theorem 3.10 above on thin triangles, if the points B and C in Fig. 3.3a are held fixed, while $y \to 0$, the defect of the triangle ABC goes to zero. The angle $\angle C$ also goes to zero, hence angle $\alpha \to \pi/2$, but the angle of parallelism is greater than α and less than $\pi/2$. Therefore $\Pi(y) \to \pi/2$ as $y \to 0$. Next, in Fig. 3.3b, let m and m' be asymptotic to ℓ (not shown) at distances y and $y + \Delta y$ along the line AB, and consider the angles labeled 1, 2, 3, 4. The angle $\angle 2$ is less than $\pi/2$, because m is asymptotic to m'. Therefore, $\angle 1 + \angle 3 > \pi/2$. Since the angle sum of the small triangle cannot exceed π, we have $\angle 3 + \angle 4 \leq \pi/2$. Subtraction gives $\angle 1 - \angle 4 > 0$. That is, $\Pi(y + \Delta y) < \Pi(y)$, so $\Pi(y)$ is decreasing. Next, to show continuity, we consider letting $\Delta y \to 0$. The angle $\angle 2 = \Pi(|PQ|) > \Pi(\Delta y) \to \pi/2$, so that $\angle 1 + \angle 3 \to \pi/2$. And by Theorem 3.10 again, $\angle 3 + \angle 4 \to \pi/2$, so that by subtraction once more, we have that $\angle 1 - \angle 4 \to 0$, hence $\Pi(y + \Delta y) \to \Pi(y)$, as required. Last, the proof that $\Pi(y) \to 0$ as $y \to \infty$ is indirect. Since $\Pi(y)$ decreases and is bounded below by 0, it has a limiting value α_0, as $y \to \infty$. We assume that $\alpha_0 > 0$

and obtain a contradiction. We consider values $\Pi(na)$ for some positive constant a and integer n. In Fig. 3.3b, we now take $y = na$ at point P and $y = (n+1)a$ at a point A (so that $\Delta y = a$). As $n \to \infty$, $\angle 4$ would $\to \alpha_0$. Now $|PQ|$ is given by the function associated with the SAA criterion in terms of side a, angle $\angle 4$, and the right angle at Q. By the continuity of that function, $|PQ|$ would converge to a positive value b, and the angle $\angle 2$ would converge to $\Pi(b)$, which is $< \pi/2$. Since the angle sum of the triangle shown is less than π, $\angle 4 < \pi/2 - \angle 3 = \angle 2 + \angle 1 - \frac{\pi}{2}$. Therefore, for large n,

$$\Pi((n+1)a) - \Pi(na) < \Pi(b) - \frac{\pi}{2}.$$

Since the right member is negative, $\Pi(na)$ would eventually become negative as n increases, and that would contradict the positivity of $\Pi(y)$. We conclude that $\alpha_0 = 0$. This completes the proof of Theorem 3.11.   □

3.4 Polar Coordinates and Ideal Points at Infinity

Let ℓ be any line and Z any point ("origin") on it, let $\overrightarrow{k}$ be one of the two rays on the line ℓ with Z as initial point, and let HP be one of the half-planes bordered by the line ℓ, called the *positive* half-plane, as in Fig. 3.4a.

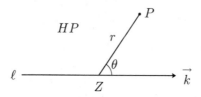

Fig. 3.4a.

Any point P in the plane can be characterized by two numbers r and θ, called *polar coordinates*, where r is the distance $|ZP|$, and θ is defined as follows: (1) it is plus or minus the measure of the angle $\angle Z\overrightarrow{P}, \overrightarrow{k}$ with the plus sign when P is in the half-plane HP and the minus sign when it is in the other half-plane, and (2) θ is 0 when P is in the ray $\overrightarrow{k}$ and is π when P is in the opposite ray $\overleftarrow{k}$. Then, θ varies in the interval $-\pi < \theta \leq \pi$. If $P = Z$, so that $r = 0$, then θ is undefined. With that understanding, there is a one-to-one correspondence between number pairs r, θ (with $r \geq 0$) and points of the hyperbolic (or, for that matter, Euclidean) plane.

Theorem 3.12: Let r and θ be polar coordinates with respect to an origin Z and a line ℓ, as above. Let m be any other directed line in the plane, with x the coordinate on it. Let $r(x)$ and $\theta(x)$ be the polar coordinates of the point $P(x)$ on the line m. Then $\theta(x)$ has a limiting value as $x \to \infty$ and any two lines are asymptotic in the direction of increasing coordinate if and only if they have the same limiting value of θ.

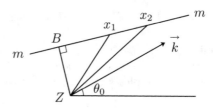

Fig. 3.4b.

Proof: Let $\overrightarrow{k}$ be the ray from the origin Z asymptotic to m, and let θ_0 be the polar angle of that ray. We consider first the case in which the line m lies as indicated in Fig. 3.4b, so that, at least for all points sufficiently far along on m, $\theta(x) > \theta_0$. Then if $x_2 > x_1$, a ray at angle $\theta(x_2)$ lies in the interior of the angle between $\overrightarrow{k}$ and the ray at angle $\theta(x_1)$; hence $\theta(x)$ is a decreasing function. Since it is also bounded below by θ_0, it has a limit as $x \to \infty$. If that limit were greater than θ_0, then any ray from the origin Z at angle θ between that limit and θ_0 would fail to intersect m, and that would contradict the statement that $\overrightarrow{k}$ is asymptotic to m. We conclude that the limit is θ_0. The other case, in which $\theta(x_1)$ and $\theta(x_2)$ are less than θ_0 (i.e., when the origin Z is above the line m in the diagram) is similar. If another line m' is asymptotic to m, then it is also asymptotic to $\overrightarrow{k}$; hence the limiting value of θ is the same on the two lines. Last, we now assume that the limiting value of θ on a line m *is* θ_0, and we wish to show that then m and $\overrightarrow{k}$ are asymptotic. We drop a perpendicular from the origin Z to m and we let B be the foot of that perpendicular. According to Section 3.1, the angle of parallelism α for the distance $|ZB|$ is the least upper bound of the angles $\angle BZX$ as the point X goes to infinity along the line m. Under the assumption we have made, that least upper bound is the angle $\theta(B) - \theta_0 = \angle ZB, \overrightarrow{k}$; hence m and $\overrightarrow{k}$ are asymptotic. $\square$

All lines of an asymptotic pencil have the same limiting value of θ. We now introduce the notion of *ideal points* ("at infinity") in the hyperbolic plane. We think of each asymptotic pencil as determining an ideal point in the direction of the lines of the pencil, and, when polar coordinates have been defined, we assign to the ideal point the polar angle θ that is associated with the pencil in the manner stated in the theorem. Since, as we shall see below, the distance between two lines of the pencil approaches zero in the limit, we think of the lines as "meeting" at the ideal point. The formal definitions are these. Each *ideal point* is an asymptotic pencil and two directed lines *meet at the same ideal point* if they are asymptotic. (We should point out that the association of values of θ with ideal points depends in a rather complicated way on the choice of the polar coordinate system.) We shall sometimes represent ideal points by symbols $A, B, \ldots$, as for ordinary points, and rely on the text for clarification.

Theorem 3.13: Any two ideal points determine a unique line; that is, there exists a unique directed line that meets one of the ideal points in the specified direction and, with direction reversed, meets the other ideal point in the reversed direction.

Proof: Let Z be the origin of a polar coordinate system, and let the ideal points have polar angles θ_1 and θ_2 with respect to that system, and with $\theta_2 > \theta_1$. Let $\vec{k}$ and $\vec{j}$ be rays from Z in those directions, as in Fig. 3.4c.

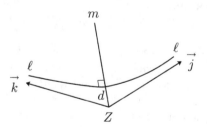

Fig. 3.4c.

We first consider the case in which $\theta_2 - \theta_1 < \pi$. (If $\theta_2 - \theta_1 = \pi$, those two rays lie on one line, and that line is the desired one. How to modify the argument given below if $\theta_2 - \theta_1 > \pi$ is left to the reader.) Let m be a line through Z at angle $\theta = 1/2 \, (\theta_1 + \theta_2)$, as shown; m bisects the angle $\angle \vec{j}, \vec{k}$ and makes an angle $\alpha = 1/2 \, (\theta_2 - \theta_1)$ with each of the rays. Since $\alpha < \pi/2$ and since the angle of parallelism $\Pi(y)$ decreases continuously from $\pi/2$ to 0 as y varies from 0 to infinity, there is a unique distance d such that $\Pi(d) = \alpha$. The desired line ℓ is the line perpendicular to m and intersecting m at distance d from Z, as shown in Fig. 3.4c. Since α is the angle of parallelism for the distance d, ℓ is asymptotic to $\vec{j}$ in one direction and to $\vec{k}$ in the other. $\qquad\square$

Since the case of one finite point and one ideal point can be taken care of by observing that a line can be drawn from any point in any direction, we see that the statement in Axiom 1 can be generalized to say that *any* two points, finite or ideal, determine a unique line. (That is not true in the Euclidean case.)

Exercise

1.   Given an asymptotic pencil and a line ℓ not in it, show that there is a unique line of the pencil perpendicular to ℓ.

3.5 Ultraparallel Lines

If two lines do not intersect and are not asymptotic in either direction, they are called *ultraparallel*. (In this section, we are concerned only with the hyperbolic case.) For example, if two lines are both perpendicular to a third line, they do not intersect, according to the alternate interior angle theorem, and they are not asymptotic because the angle of parallelism is always less than 90°. We shall see that, conversely, any two ultraparallel lines do have a common perpendicular, and the common perpendicular is unique, for the given lines. We use properties of Lambert quadrilaterals. (Recall that a Lambert quadrilateral is one with three right angles.)

Remark: Examination of the proof of Theorem 2.13 in Section 2.8 shows that in the hyperbolic plane, where every polygon has a nonzero defect, that theorem can be strengthened. In a Lambert quadrilateral in the hyperbolic plane, the fourth angle is *less than* 90° and each side adjacent to that angle is *longer than* the side opposite to it.

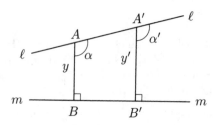

Fig. 3.5a.

Let ℓ and m be nonintersecting lines, with x a coordinate on ℓ. From point A on ℓ with coordinate x, we drop a perpendicular to m and we denote by $y = y(x)$ the length $|AB|$ of that perpendicular, as shown in Fig. 3.5a. One of the angles between ℓ and the perpendicular at A is greater than or equal to $\pi/2$, and we assume that it is the angle α on the side of increasing coordinate x on ℓ (in the opposite case, we reverse the direction of ℓ by replacing x by $-x$). Let A' be a point with coordinate $x' > x$. Since the angular defect of the quadrilateral $ABB'A'$ is positive, the angle $\alpha' = \alpha(x')$ is greater than $\alpha(x)$. By comparing that quadrilateral with the Lambert quadrilateral having right angles at B and B' and also at A (this right angle is not shown in the figure), we see that $y(x') > y(x)$. Therefore, $\alpha(x)$ and $y(x)$ are increasing functions, at least for all $x > x(A)$, where A is the point we started with. As x *decreases* from $x(A)$, those functions may decrease indefinitely, or $\alpha(x)$ may become 90° at some point A_0. Then, $y(x)$ increases again with further decrease of x. At A_0, $y(x)$ has a minimum and the lines have a common perpendicular. The case in which the lines have no common perpendicular is discussed more fully below.

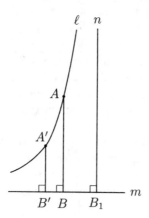

Fig. 3.5b.

We now show that as $x \to +\infty$, $y(x)$ also $\to +\infty$. According to Exercise 1 in the preceding section, there is a line n asymptotic to ℓ and perpendicular to m at some point, say B_1, as in Fig. 3.5b. Intuitively, the perpendicular to m dropped from the ideal point at the end of ℓ is of infinite length; hence $y(x)$ would seem to become arbitrarily large as x increases. If you want a formal proof, do Exercise 2 below.

We now consider the case in which $y(x)$ has no minimum. Since $y(x)$ decreases as $x \to -\infty$, but is positive (we assumed that the lines are non-intersecting), it has a limit y_1, and we claim that $y_1 = 0$. We assume that the limit y_1 is positive and show that we could then find a quadrilateral with zero angular defect, which would contradict Section 3.2. Similarly, $\alpha(x)$ would decrease to a limiting value α_1 as $x \to -\infty$. We think of moving the quadrilateral $ABB'A'$ of Fig. 3.5a indefinitely to the left, keeping the angles at B and B' at $90°$, and keeping the distance $|BB'|$ fixed at some value L. According to Exercise 1 below, if w is a coordinate on m that also decreases to the left, and if $w(x)$ is the coordinate of the foot of the perpendicular dropped from point at x on ℓ, then $w(x)$ decreases, as x decreases. It follows that if A is at any x less than the coordinate of A in the figure, then there is an x' such that $w(x') - w(x) = L$. Furthermore, $x' \to -\infty$ as $x \to -\infty$, because, by the triangle inequality, $x' - x$ cannot be more than $y(x) + L + y(x')$. The interior angle at A' of the moving quadrilateral would have the limiting value $\pi - \alpha_1$; hence the angular defect of the quadrilateral would have the limit 0 as $x \to -\infty$. We now consider a Saccheri quadrilateral with base of length L, right angles at each end of that base, and vertical sides both of length y_1. It would fit inside all the moving quadrilaterals that we have considered; hence it would have to have angular defect zero, and that gives the contradiction. Hence, $y_1 = 0$.

If the lines are asymptotic, $y(x)$ cannot have a minimum, since a common perpendicular would imply an angle of parallelism of $90°$. Therefore, for asymptotic lines, $y(x) \to 0$ at one end and ∞ at the other. Conversely, if $y(x)$ goes to zero at one end, then the lines are asymptotic, for it then follows from the corollary to Theorem 2.13 in Section 2.8 that, when a polar coordinate system is established, the two lines have the same limiting value of the polar angle θ, and then it follows from Theorem 3.12 in Section 3.4 that the lines are asymptotic. We summarize.

Theorem 3.14: Ultraparallel lines have a unique common perpendicular and the distance between them increases without limit in either direction from the common perpendicular. The distance between asymptotic lines converges to 0 in the direction in which they are asymptotic and increases without limit in the other direction.

Exercises

1. Explain what one means by saying that if ℓ and m are nonintersecting, a coordinate x on ℓ and a coordinate w on m increase *in the same direction*. *Hint*: Consider a transversal. Then show that if $w(x)$ is the coordinate of the foot of a perpendicular dropped to m from point x on ℓ, then $w(x)$ is an increasing function. *Hint*: If $x < x'$, why cannot $w(x)$ be equal to $w(x')$, and why cannot it be greater than $w(x')$?

2. Suppose that line n is asymptotic to ℓ and perpendicular to m at point B_1, as in Fig. 3.5b. Show that as $x \to \infty$, the point B at $w(x)$ approaches B_1 and $y(x) \to \infty$, where $y(x)$ is the distance from m of a point on ℓ at coordinate x. *Hint*: Let A' be a fixed point on ℓ at coordinate x'. Then let A be at $x > x'$, and let B' and B be the feet of the corresponding perpendiculars. To show that $|BB_1| \to 0$, drop a perpendicular from A to n and use a property of Lambert quadrilaterals.

To show that $|AB| \to \infty$, show by the triangle inequality that

$$|AB| > x - x' - |A'B| > x - x' - |A'B_1|.$$

3. Let ℓ and m be nonintersecting lines, with ideal endpoints at A, A' for ℓ and B, B' for m. Discuss the lines from A to B' and from A' to B. Do they intersect? If so, where?

4. Show that the figure consisting of two ultraparallel lines is invariant under reflection in their common perpendicular. (See the next section for an analysis of reflections.)

3.6 Isometries

In Section 10 of Chapter 2, we introduced the isometry group of $\mathbb{P}$ as one expression of the homogeneity and isotropy of the plane. Although Chapter 3 is primarily about $\mathbb{H}^2$, until later in this section, all considerations apply equally to the Euclidean plane $\mathbb{E}^2$ and the hyperbolic plane $\mathbb{H}^2$. The symbol $\mathbb{P}$ may represent either of them. We continue to write $\mathbb{G}$ for the isometry group.

We begin by showing how to produce isometries in the plane using two independent sets of polar coordinates r, θ and r', θ'. The first set is determined as in Fig. 3.4a of Section 3.4, and the second is determined as in Fig. 3.6a, in terms of another choice of origin Z', line ℓ', ray $\overrightarrow{k}'$, and half-plane HP'. We introduce a mapping σ of the plane onto itself by mapping a point P with coordinates r, θ onto the point P' with the same coordinate values in the primed system: $r' = r$ and $\theta' = \theta$.

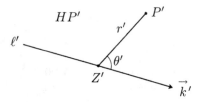

Fig. 3.6a.

We show that σ is an isometry. If P and Q are two points and P' and Q' are their images, then, first, the angles $\angle PZQ$ and $\angle P'Z'Q'$, with vertices at the corresponding origins, are equal because they are both equal to $\theta(P) - \theta(Q)$, and it follows that $|PQ|$ is equal to $|P'Q'|$ by the SAS criterion. That is, the mapping σ preserves distances. Now, P can be any point and then P' is uniquely determined. Conversely, P' can be any point, and then its preimage P is uniquely determined; hence σ maps the entire plane one-to-one onto itself and is an isometry as defined above.

Any isometry of the plane can be produced in that way. It is only necessary to choose $Z, \ell, \vec{k}, HP$ arbitrarily, and then set $Z' = \sigma Z$, $\ell' = \sigma \ell$, $\vec{k}' = \sigma \vec{k}$, and $HP' = \sigma HP$, where $\sigma \ell$ denotes the line consisting of all the images of the points in ℓ, and so on.

We now consider special cases. First, assume that ℓ and ℓ' are the same line (while Z and Z' are different points on it) and the rays $\vec{k}$ and $\vec{k}'$ are in the same direction on that line and the half-planes HP and HP' are the same. Then the isometry is called a *translation along the line ℓ by a distance* $X = x(Z') - x(Z)$, where x is a coordinate on ℓ in the direction of $\vec{k}$. That translation is denoted by T_X.

Next let Z and Z' be the same point, while $\vec{k}$ and $\vec{k}'$ are in different lines through Z and make an angle α. The mapping is a rotation or a rotation combined with a reflection (but see Exercise 3 below), depending on the relation of the two half-planes. The cases that lead to a rotation are as follows: (1) The new ray $\vec{k}'$ is in the old half-plane HP, and HP' is the half-plane bounded by $\vec{k}'$ that does not contain $\vec{k}$, as in Fig. 3.6b.

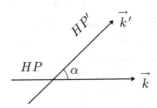

Fig. 3.6b.

Detailed consideration shows that if P is a point with angular coordinate $\theta = \theta(P)$ and P' is its image, then the coordinate of P' in the unprimed system is

$$\theta(P') = \theta(P) + \alpha \pmod{2\pi}. \tag{3.6-1}$$

(2) The new ray $\vec{k}'$ is in the half-plane opposite to HP, and HP' is the half-plane bounded by $\vec{k}'$ that does contain $\vec{k}$. The above equation holds. (3) $\vec{k}' = \vec{k}$ and $HP' = HP$; then $\alpha = 0$. (4) $\vec{k}'$ and $\vec{k}$ are opposite and HP' and HP are opposite; then $\alpha = 180°$.

Last, if Z and Z' are the same and the rays $\vec{k}$ and $\vec{k}'$ are the same, while HP and HP' are opposite, the mapping is called a *reflection* in the line ℓ and is denoted by M (for "mirror") or by M_ℓ. That amounts to mapping point P at r, θ onto P' at $r, -\theta$. Since PZP' is isosceles, P' is obtained by dropping a perpendicular from P to ℓ and continuing to an equal distance on the other side of ℓ.

We now show that a general isometry can be represented as the composition of mappings of the above three types. Consider the isometry σ described above as a mapping from the r, θ polar coordinate system to the r', θ' system. First, we can "undo" the mapping (find its inverse) by first performing a rotation R_{α_1} to bring the origin Z' onto the axis of $\theta = 0$ (with $\alpha_1 = 0$ if Z' was already there); then we perform a translation T_X along ℓ that brings Z' into coincidence with Z, then a rotation R_{α_2} about the origin Z that brings $\overrightarrow{k}'$ into coincidence with $\overrightarrow{k}$. Then the half-plane HP' is either the same as HP or opposite; in the latter case, we perform a reflection M. Since the inverse of a general isometry is a completely general isometry, we have proved the following:

Theorem 3.15: The general isometry can be written either as

$$\sigma = R_{\alpha_2} T_X R_{\alpha_1} \text{ or as } \sigma = M R_{\alpha_2} T_X R_{\alpha_1}. \tag{3.6-2}$$

So far, we have defined translations only along the line ℓ and rotations only about Z. Other translations and other rotations can be defined in evident ways. For example, for any α and any X, $R_{-\alpha} T_X R_\alpha$ is a translation along some other line through the origin, and $T_{-X} R_\alpha T_X$ is a rotation about another point on the line ℓ. (But see Exercise 4 below.)

Isometries that contain no reflection, as on the left of (3.6-2), are *direct* because they preserve orientation; ones that contain a single reflection, as on the right, are called *reverse* because they reverse orientation. The structures of direct and reverse isometries in the hyperbolic plane are discussed in Section 3.8 below.

The Euclidean and hyperbolic planes have quite different isometry groups, because the individual isometries combine in different ways. The following are some of the differences.

(1) In the Euclidean plane, a translation by a certain distance along a line ℓ is exactly the same mapping as the translation by the same distance along any line m parallel to ℓ. In the hyperbolic plane there are many lines m parallel to ℓ through any point P, and translations along those lines m are all different; hence we should expect that there are more translations, in a sense, in the hyperbolic case than in the Euclidean. It will be seen in Chapter 8 that although a Euclidean translation can be specified by giving the values of two parameters, three parameters are required in the hyperbolic case.

(2) In the Euclidean case, the composition of two translations is another translation (obtained by vector addition of the 2-component vectors that represent the translations). As shown in the next section, the composition

of two translations in the hyperbolic plane is not generally a translation at all. Hence, the set of all translations form a subgroup of $\mathbb{G}$ in the Euclidean case but not in the hyperbolic case.

(3) In the Euclidean case, any direct isometry has either a fixed point (i.e., a fixed *finite* point) (in which case it is a rotation about that point) or an invariant line (in which case it is a translation along that line). In the hyperbolic plane, an isometry may not have either a fixed point or an invariant line. It is then neither a rotation about any point nor a translation along any line, but a new kind of isometry, called an *ideal rotation*. See Section 3.8.

We are thus led to the following problem.

Problem: Given an isometry of the hyperbolic plane, determine whether it has a fixed point or an invariant line. (A line is called *invariant* if every point of it is mapped onto a point of it.) If it has neither, describe the isometry. The solution of this problem, in terms of the parameters appearing in (3.6-2) is found in Section 3.8 below. Its solution in terms of matrix representations of the isometries is found in Chapter 8.

Exercises

1. Show that if an isometry has a fixed point Z, it is either the identity or a rotation about the point Z or a reflection in a line that passes through the point Z.
2. Show that if an isometry leaves a line invariant, then it is either a translation along that line, or such a translation followed by a reflection in the line.
3. Show that if a rotation R_α about the origin is combined, in either order, with the reflection M described above, the result is a reflection in some line through the origin.
4. Show that a rotation about a point in the line ℓ can be written in the first form in (3.6-2). In that case, what is the relation between the angles α_1 and α_2?
5. Let $\omega_1, \omega_2, \omega_3$ be any three distinct ideal points at infinity, and let $\omega_1', \omega_2', \omega_3'$ also be any three distinct ideal points at infinity. Under what circumstances does there exist a direct isometry of the plane that maps ω_i onto ω_i' ($i = 1, 2, 3$), and under what circumstances does there exist a reverse isometry that does the same?
6. Under what circumstances are two quadrilaterals with all four vertices at infinity congruent?
7. Under what circumstances are two triangles, each with two vertices at infinity, congruent?
8. Under what circumstances are two generalized quadrilaterals with three right angles and the fourth vertex at infinity congruent?

3.7 Rotation by a Composition of Translations

We now show that, in the hyperbolic plane, the composition of three pure translations can be a rotation. (A similar composition of two translations is considered in Exercise 1 below.) Let $\triangle ABC$ be a given triangle, with angles α, β, γ at A, B, C, respectively. Consider three translations as follows (see Fig. 3.7a):

$$\sigma_1 = \text{translation by distance } |AB| \text{ along the ray } \overrightarrow{AB},$$

$$\sigma_2 = \text{translation by distance } |BC| \text{ along the ray } \overrightarrow{BC},$$

$$\sigma_3 = \text{translation by distance } |CA| \text{ along the ray } \overrightarrow{CA}.$$

In the composition of these mappings, taken in the above order, the point A is mapped onto itself. The composition preserves orientation so it is either the identity or a rotation about A, and we shall show that it is in fact a rotation through the angle $\pi - \alpha - \beta - \gamma$, which is the angular defect of the triangle, hence is $\neq 0$.

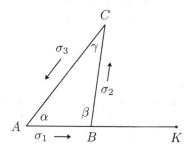

Fig. 3.7a.

The ray $\overrightarrow{AB}$ is mapped by σ_1 onto a ray $\overrightarrow{BK}$ on the continuation of $\overrightarrow{AB}$, as shown in the drawing. We write $\overrightarrow{BK} = \sigma_1(\overrightarrow{AB})$, and so on. By analysis of the angles, we find that

The angle between $\sigma_1(\overrightarrow{AB})$ and $\overrightarrow{BC}$ is $\pi - \beta$.

The angle between $\sigma_2(\sigma_1(\overrightarrow{AB}))$ and $\overrightarrow{AC}$ is $\pi - \beta - \gamma$.   (3.7-1)

The angle between $\sigma_3(\sigma_2(\sigma_1(\overrightarrow{AB})))$ and $\overrightarrow{AB}$ is $\pi - \beta - \gamma - \alpha$.

That is, the ray $\overrightarrow{AB}$ has been rotated through the angle $\pi - \beta - \gamma - \alpha$. We say that the ray $\overrightarrow{AB}$ has been *carried round the triangle by parallel transport.* Parallel transport round any closed polygon can be similarly defined. It produces a rotation equal to the angular defect of the polygon. In differential geometry there is further generalization to the parallel transport round any closed curve in a general Riemannian space.

Exercises

1. Let $ABCD$ be a so-called Saccheri quadrilateral with right angles at A and B and with $|AD| = |BC|$, as in Fig. 3.7b. (This is not a rectangle; there are no rectangles in the hyperbolic plane.) Let σ_1 be the translation along $\overrightarrow{AB}$ by the distance $|AB|$, and let σ_2 be the translation along $\overrightarrow{CD}$ by the distance $|CD|$. Discuss the composition $\sigma_2 \circ \sigma_1$ of these two mappings.

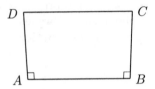

Fig. 3.7b.

2. Show the steps involved in establishing the relations (3.7-1).

3.8 Analysis of Isometries by Reflections

The difference between direct and reverse isometries can be characterized as follows. First, we think of a polar coordinate system r, θ as referring to points in a diagram drawn on a piece of paper, and we call the system *clockwise* or *counter-clockwise* according as an increase of θ within the interval $(-\pi, \pi)$ at constant r causes the point r, θ to move clockwise or counter-clockwise about the origin, as we view the paper from above. When an isometry is given in terms of two coordinate systems r, θ and r', θ', as in Section 3.6, the isometry is *direct* if both coordinate systems are clockwise or both counter-clockwise. Otherwise it is *reverse*. It is easily verified that a rotation R_α and a translation T_X are both direct, while a reflection M is a reverse isometry. When an isometry is the resultant of any finite collection of mappings, each of which is a rotation R_α or a translation T_X or a reflection M, the resulting isometry is direct or reverse according to whether the number of reflections M in the collection is even or odd.

We showed in Section 10 of Chapter 2 that every direct isometry is a composition of two reflections, say in lines ℓ and n. If the lines ℓ and n intersect in a single point, the isometry is a rotation about the point of intersection. If they are asymptotic, it is an ideal rotation, which will be described below. If they are ultraparallel, it is a translation along the line that is the common perpendicular of the lines. If the two lines are identical, the isometry is the identity.

We now consider the case of ideal rotation in more detail. We assume that the lines ℓ and n have had directions put on them and are asymptotic in that direction, as defined in Section 3.1. The reflection M_ℓ, being an isometry, maps asymptotic lines into asymptotic lines. In particular, it maps any line of the asymptotic pencil to which it belongs onto another line of that pencil. M_n does the same, because it belongs to the same pencil. Therefore, the direct isometry $\sigma = M_n M_\ell$ maps each line of the pencil onto another line of the pencil. Since the pencil represents an ideal point at infinity, the mapping is a direct isometry that leaves fixed the ideal point at infinity at which the lines ℓ and n meet. It may be thought of as a "rotation" about that ideal point. It is sometimes called a *parallel displacement* (because some authors, starting with Lobachevski, have used the word "parallel" for our "asymptotic").

In the next section, an explicit formula for an ideal rotation is found, in terms of a special set of coordinates in the plane.

Exercises

1. When the lines ℓ and n intersect at a finite point, determine the angle α of the resulting rotation.
2. When the lines ℓ and n are ultraparallel, find the distance X of the resulting translation.
3. Let T_X and R_α be isometries as defined in Section 3.6 in reference to a particular polar coordinate system. Let y be a positive distance. Let $\alpha = \Pi(y)$ and $\beta = 90° - \alpha$. Show that the isometry

$$\sigma = T_y R_{2\beta} T_y \qquad (3.8\text{-}1)$$

    is an ideal rotation. Where is the ideal point invariant under it?
4. Show that no finite point is fixed under an ideal rotation. (It is assumed, of course, that ℓ and n are not the same line.)
5. Show that no line is invariant under an ideal rotation. *Hint*: If P is any point, $P' = \sigma P$, and $P'' = \sigma P'$, show that P, P', P'' are not collinear.

Now let σ be a reverse isometry. If σ has any fixed points, it must have at least a line of fixed points, because an isometry with exactly one fixed point is a rotation. An isometry with a line of fixed points is either a reflection or the identity. Since the identity is direct, if σ has any fixed points it must be a reflection. We will prove that a reverse isometry σ that

has no fixed points is what is called a *glide reflection*. That is, there exists
a line ℓ and a number x such that

$$\sigma = M_\ell T_{\ell,x}, \qquad\qquad (3.8\text{-}2)$$

so that σ is obtained by translating along ℓ and then reflecting in ℓ.

One feature of a glide reflection is that it exchanges the half-planes on
opposite sides of ℓ. If Z is not on ℓ, let F be the foot of the perpendicular
from Z to ℓ. Let F' be the translate of F along ℓ, which is its image under
the glide reflection. Then the image of Z under the glide reflection is the
point in the half-plane opposite Z whose distance from F' is $|ZF|$. Thus
the midpoint between a point and its image is always on ℓ. We can use
this fact to find the glide reflection representation of an arbitrary reverse
isometry that has no fixed points.

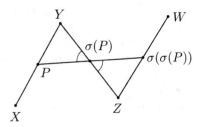

Fig. 3.8.

Let σ be a reverse isometry that has no fixed points. We seek a point
P such that the points P, $Q = \sigma(P)$, and $R = \sigma(\sigma(P))$ are collinear. If
we have such a point, then P, Q, and R must be distinct: if not, $R = P$
(why?), and the midpoint between P and Q is fixed by σ, a contradiction.
Let ℓ be the line through P and Q, and let τ be the translation along ℓ that
takes P to Q. Then τ also takes Q to R. Hence, $\sigma \circ \tau^{-1}$ fixes every point
of ℓ, and being a reverse isometry, it must be M_ℓ.

To find such a point P, start with any point X, let Y be its image
under σ, let Z be the image of Y and let W be the image of Z, as in
Fig. 3.8. Let P be the midpoint of the segment XY. Then the image of P
is the midpoint of YZ and its image is the midpoint of ZW. The image of
the segment $P\sigma(P)$ is the segment $\sigma(P)\sigma(\sigma(P))$. Because σ is an isometry,
the triangles $PY\sigma(P)$ and $\sigma(P)Z\sigma(\sigma(P))$ are congruent. Hence the angles
$\angle Y\sigma(P)P$ and $Z\sigma(P)\sigma(\sigma(P))$ are congruent. It follows that we have three
collinear points, as desired.

Further Exercises

6. Is the σ given by (3.8-2) also equal to $T_{\ell,X}M_\ell$?
7. Represent the mapping σ given by (3.8-2) by a composition of three reflections.
8. Let σ be the resultant of a rotation about a point Z followed by a reflection in a line m that does not pass through Z. Show how to find the line ℓ and the distance X when σ is represented in the form (3.8-2).
9. Show that given any two lines of an asymptotic pencil, there is an ideal rotation that carries one into the other. *Hint*: See Fig. 3.1g.
10. Let m and n be lines at right angles. Let σ be an ideal rotation about the ideal point at one end of m followed by a reflection in n. Show how to determine ℓ and X when σ is written as (3.8-2). *Hint*: Write σ as a composition of three reflections, and then recombine them in a different order.
11. Show that the set of all ideal rotations about a given ideal point at infinity constitutes a group (a subgroup of the isometry group).

3.9 Horocycles — A Special Coordinate System

Let P,Q be points on a line ℓ, and let C be a circle of radius $|PQ|$ with center at Q. In the Euclidean case, if the point Q moves off to infinity long the line ℓ, keeping P fixed, the circle converges to the straight line through P perpendicular to ℓ in the limit. In the hyperbolic case, it converges to a curve called a *horocycle*. Let PR be a chord of the circle, with midpoint at S as in Fig. 3.9a. Then the segment QS is perpendicular to the chord PR. Let r,θ be polar coordinates of the point R, with P as origin and the line ℓ as the axis of $\theta = 0$. Then $|PR| = r$, $|PS| = r/2$. We let Q move to the right, but we keep r fixed, so that the point R moves on a circle of radius r about the point P. Since the baselength of the isosceles triangle PRQ is fixed, the angle θ increases, but remains less than $90°$, and we wish to find its limiting value. The angle $\angle PSQ$ continues to be a right angle, and the lines PQ, SQ, and RQ become asymptotic as Q moves to infinity along the line ℓ, so that the angle $\theta = \angle SPQ$ becomes equal to the angle of parallelism $\Pi(y)$, for $y = |SP| = r/2$. Hence, in the limit

$$\theta = \Pi\left(\frac{r}{2}\right). \tag{3.9-1}$$

We now let r (hence also θ) vary. The above relation between them is the equation in polar coordinates of a curve called a *horocycle*; an explicit equation for it will be found in Chapter 6. (We have as yet no formula for the function Π.)

For any r, the final configuration is as in Fig. 3.9b, where the lines ℓ,m,n are all asymptotic. R is the image of P under reflection in the line m, and the horocycle consists of the images of P under reflection in all the lines of the asymptotic pencil.

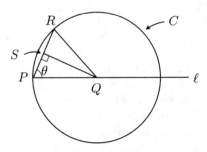

Fig. 3.9a.

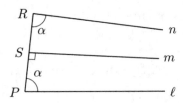

Fig. 3.9b.

Furthermore, R is then invariant under reflection in n, hence the horocycle can also be characterized as the images of P under ideal rotations about the ideal point where those lines meet. According to Exercise 11 in the preceding section, those ideal rotations form a group; hence, if R and R' are any two points of the horocycle, each of them is the image of P under a member of that group; hence R' is the image of R under a member of the group. That is, the entire horocycle is invariant (mapped onto itself) by ideal rotations. At P, the horocycle lies between any of the circles mentioned and their tangent at P; hence the horocycle is perpendicular at P to the line ℓ. Hence, since the ideal rotations are isometries, the horocycle cuts all the lines of the asymptotic pencil orthogonally. If we repeat the entire construction starting with another line ℓ' and point P', then the entire Fig. 3.9a is replaced by another figure congruent to it, with $|PQ| = |P'Q'|$, etc. Hence, all horocycles are congruent. In particular, if ℓ and ℓ' are the same line, we get in this way a new horocycle obtained from the first by a translation through the distance $|PP'|$ along the line ℓ. Each line of the asymptotic pencil cuts both horocycles orthogonally, and since the entire structure is invariant under ideal rotations, we see that the distance between the horocycles along such a line is equal to $|PP'|$ for all lines of the pencil. (See Exercise 6 below.)

We now introduce the notion of *arclength* along a horocycle. Generally, if C is any piece of curve given parametrically by $P(t)$ for $a \leq t \leq b$, where, for each t, $P(t)$ is a point on C, we consider a partition $\mathcal{P}$ of the interval $[a, b]$ by values t_i of the parameter such that

$$\mathcal{P} : a = t_0 < t_1 < \cdots < t_n = b.$$

Then a polygonal approximation to the length $L(C)$ is given by

$$s(\mathcal{P}) = \sum_{i=1}^{n} |P(t_{i-1})P(t_i)|,$$

where, as usual, $|AB|$ denotes the distance from A to B. We consider a sequence of partitions $\mathcal{P}_k$, where $\mathcal{P}_k$ has n_k points in the interval $(a, b]$, and $n_k \to \infty$ as $k \to \infty$. For simplicity, we assume that $\mathcal{P}_{k+1}$ contains all the values t_i of $\mathcal{P}_k$ and more too. It then follows from the triangle inequality that the approximations increase increase as $k \to \infty$. We also assume that the largest increment $\text{Max}_{(i)}(t_i - t_{i-1})$ tends to zero, as k goes to infinity. If we can somehow prove that those approximations are bounded above for all k, we say that the curve C is *rectifiable*, and we define

$$L(C) = \lim_{k \to \infty} s(\mathcal{P}_k). \tag{3.9-2}$$

In particular, let C be a portion of the horocycle given by (3.9-1), with the parameter t taken as r. (Recall from Theorem 3.11 in Section 3.3 that $\Pi(y)$ is a continuous decreasing function of y.) Then $L(C)$ is the *arclength* of the part of the horocycle for $a \leq r \leq b$ (where we assume that a is nonnegative — see Exercise 8 below). If $b - a$ is not too large (see Exercise 7 below), we can find an upper bound for the polygonal approximations by placing a tangent line to the horocycle at some point of the part C. Then the length L_0 of that part of the tangent line cut out by the lines of the asymptotic pencil through the endpoints of C at $r = a$ and $r = b$ is an upper bound for the polygonal approximations. (For larger values of $b - a$, the arclength can be obtained by adding the arclengths of shorter pieces.)

We now describe a coordinate system in which there are explicit formulas for the isometries in a certain subgroup of the isometry group.

We consider an asymptotic pencil of lines (Section 3.1) and the family of horocycles everywhere orthogonal to those lines, as shown in Fig. 3.9c. With respect to an arbitrary origin O, we define coordinates ξ, η as follows: η is constant along each line; it is the arclength from the point O along the primary horocycle (the one that passes through O), and ξ is the distance along any of the lines, taken as zero at the primary horocycle and increasing to the left. Then, ξ and η are orthogonal coordinates that cover the entire hyperbolic plane; each of them varies from $-\infty$ to $+\infty$.

We now consider arclength s along a horocycle at position $\xi = \xi_1$, and we denote by $s(\xi_1, \eta)$ the arclength along that horocycle from the point $\xi_1, 0$

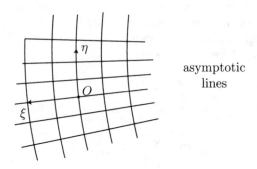

Fig. 3.9c.

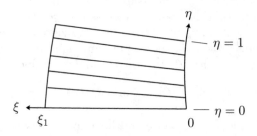

Fig. 3.9d.

to the point ξ_1, η. We shall show that for fixed $\xi_1, s(\xi_1, \eta)$ is proportional to $\eta = s(0, \eta)$. We consider the region bounded by lines at $\eta = 0$ and $\eta = 1$ and by horocycles at $\xi = 0$ and $\xi = \xi_1$, as in Fig. 3.9d.

Let the interval $[0, 1]$ of η be divided into q equal subintervals to give lines at $\eta = p/q$, $(p = 0, 1, 2, \ldots, q)$. As we have seen, there is an ideal rotation or "rotation about a point at infinity" that maps each line of the pencil onto another line of the pencil, leaving the horocycles invariant. The ideal rotation that maps the line at $\eta = 0$ onto the line at $\eta = 1/q$ maps each of the lines in the figure onto the one just above it. Since it is an isomorphism, all the thin strips in the figure are congruent; hence those lines also subdivide the horocycle arc at $\xi = \xi_1$ into equal subarcs. It follows that

$$s\left(\xi_1, \frac{p}{q}\right) = \frac{p}{q}\, s(\xi_1, 1)$$

Since the function $s(\xi, \eta)$ is clearly continuous in each of its variables, the above holds for irrational values of η, and we have generally

$$s(\xi, \eta) = \eta s(\xi, 1).$$

Now consider a second figure congruent to the one in Fig. 3.9d obtained by a translation by a distance ξ_2 along the ξ-axis, as shown in Fig. 3.9e.

Fig. 3.9e.

Since translation is an isomorphism,* the amplification of the horocycle arclength from $\xi = 0$ to $\xi = \xi_1$ is the same as that from ξ_2 to $\xi_2 + \xi_1$, hence

$$\frac{s(\xi_2 + \xi_1, 1)}{s(\xi_2, 1)} = \frac{s(\xi_1, 1)}{1},$$

so that

$$\ln s(\xi_2 + \xi_1, 1) = \ln s(\xi_2, 1) + \ln s(\xi_1, 1).$$

We know from calculus that if a continuous function $f(x)$ is such that $f(a+b) = f(a) + f(b)$, then $f(x) = \text{const} \cdot x$. We conclude that $s(\xi, 1) = e^{K\xi}$, where K is a constant, hence

$$s(\xi, \eta) = \eta e^{K\xi}. \tag{3.9-3}$$

Clearly, K is positive, because the above quantity increases with increasing ξ. We have proved the following:

Lemma 3.2: Let ℓ and ℓ' be two lines of the asymptotic pencil. The distance (arclength) along a horocycle at position ξ from ℓ to ℓ' is proportional to $e^{K\xi}$, as ξ varies, for fixed ℓ and ℓ', where K is a constant.

The value of K clearly depends on the unit of length in the hyperbolic plane, which has so far not been chosen (see Section 1.2). (Often, that unit of length will be so chosen that $K = 1$.)

Equation (3.9-3) is the first quantitative formula of the theory; other formulas will be obtained in Chapter 6. The formula was obtained by appeal to the isometries of the hyperbolic plane. That is a special case of the thing pointed out in 1872 by F. Klein in his "Erlanger Program," according to which the geometry of a space is largely determined by its isometry group.

The isometries that we have used are ideal rotations R_A and translations T_X along the ξ-axis (the line at $\eta = 0$). Together, these generate a

---

* Recall: All horocycles are congruent.

two-parameter subgroup of the isometry group. The corresponding mappings are:

$$R_A : \xi \to \xi, \qquad \eta \to \eta + A, \tag{3.9-4}$$
$$T_X : \xi \to \xi + X, \qquad \eta \to e^{-KX}\eta. \tag{3.9-5}$$

It can be shown (Exercise 2) that a commutator

$$T_X R_A T_{-X} R_{-A} \tag{3.9-6}$$

is of the form $R_{A'}$ for some A'. Therefore, in any product of the form $T_1 R_1 T_2 R_2 T_3 R_3 \cdots$, the R_i can be moved to the left in the product and the T_i to the right. The result is of the form $R_A T_X$ for some A and some X; hence the subgroup generated has two parameters, A and X.

Exercises

1. Verify (3.9-4) and (3.9-5).
2. Show that the commutator (3.9-6) is of the form $R_{A'}$.
3. Show that $R_A T_X R_{-A}$ is a translation along another line of the pencil.
4. Show that under any transformation of the subgroup a vector that points to the ideal point at infinity (of the pencil) continues to do so afterward.
5. For what reflections can you give formulas in terms of the coordinates ξ and η?
6. Resolve the following apparent intuitive contradiction: It was said that the horocycle through B can be obtained from the one through A by a translation by the distance $|AB|$ along the line ℓ. Under that translation, the point C moves to the point D in Fig. 3.9f. It would seem then that the *shortest* distance from the point C to the second horocycle, namely, the distance $|CE|$ along the common perpendicular would be less than $|AB|$, not equal to $|AB|$, as stated in the text. *Hint:* See the next section.

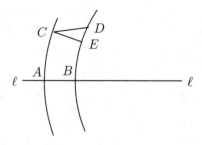

Fig. 3.9f.

7. Explain why, in the discussion of the upper bound L_0 for the polygonal approximations by means of a tangent line to a portion of the horocycle, it was necessary to assume that $b - a$ is not too large, and then show that with that restriction, L_0 *is* an upper bound.
8. In that same discussion, it was assumed that $0 \leq a < b$. What is to be done if $a < b \leq 0$ or if $a < 0 < b$?
9. Find the equation in the ξ, η system of a general equidistant to a line $\eta = \eta_0$ (a constant).

3.10 Equidistants

The circles and the horocycle constitute a continuous family of types of curves, each of which is invariant under a (one-parameter) subgroup of $\mathbb{G}$, so that it slides along itself under the isometries of the subgroup. When drawn so as to pass through the origin vertically, they look roughly as shown in Fig. 3.10a. They occupy the shaded region of the plane.

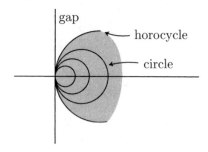

Fig. 3.10a.

The question naturally arises whether there is another family of "sliding" curves that fill the gap between the horocycle and the vertical straight line. There is, and the curves are called *equidistants*. Let m be a straight line, as shown vertically in Fig. 3.10b. For each point P define the distance (P, m) by dropping a perpendicular to m and defining that distance as $|PQ|$, where Q is the foot of the perpendicular. The set of all points P at a given distance x from m is a curve C called an *equidistant*. It is clearly invariant under translations along m, because a perpendicular PQ, when translated, becomes another perpendicular $P'Q'$ of the same length x, so C slides along itself under a translation along m. For $x = 0$, C is the line m. As x increases and the line m moves to the right to infinity, the line

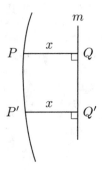

Fig. 3.10b.

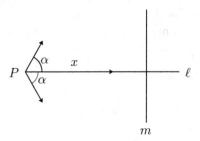

Fig. 3.10c.

m becomes a point at infinity (see Comment below), and C converges to a horocycle.

Comment: It was seen in Section 3.3 that the angle of parallelism $\alpha = \Pi(x)$ goes to zero as $x \to \infty$. Hence, if P is a fixed point on a line ℓ, and if m is a line perpendicular to ℓ at a distance x, the line m is confined to the interior of the angle of measure 2α shown in Fig. 3.10c for x large enough so that $\Pi(x) < \alpha$. As α tends to zero, the two rays shown approach coincidence, and in this sense we can say that the entire line m converges to the ideal point at infinity along the line ℓ.

3.11 Tiling, Lattices, and Triangulations

A *tiling* of a plane (Euclidean or hyperbolic) is, roughly speaking, a set of bounded regions $R_0, R_1, \ldots$, all congruent to each other, which cover the plane without overlap (which means that no two of them have common interior points — they may have common boundary points). Tilings of

the Euclidean plane by parallelograms and by hexagons are indicated in
Fig. 3.11a.

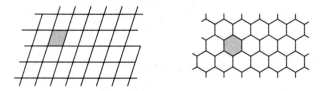

Fig. 3.11a.

The words "roughly speaking" in the above definition refer to two
provisos:

(1) We shall consider only polygonal regions, not regions with curved sides.
(2) The congruences involved must constitute a subgroup G_0 of the group
     $\mathbb{G}$ of isometries of the plane, and the regions of the tiling must be
     precisely all those regions that can be obtained from one of them (e.g.,
     the one shaded in the drawings) by the isometries of the subgroup G_0.

For the parallelogram tiling shown in the first drawing, the subgroup G_0
can be taken as the group of all translations by amounts of the form $n\vec{u} +$
$m\vec{v}$, where $\vec{u}$ and $\vec{v}$ are noncollinear vectors that describe the sides of the
parallelograms, and n and m are arbitrary integers (positive, negative, or
zero).

Note 1. In the hyperbolic case, the subgroup G_0 cannot be of the
kind just described, because, as noted in Section 3.6, a collection of pure
translations cannot be a group, unless they are all translations along the
same line.

Note 2. In the parallelogram tiling of the Euclidean plane described
above, the subgroup could be taken also to include 180° rotations about
the centers of the parallelograms, about their vertices, and about the centers
of their sides. If the parallelograms are squares, there are further isometries
that could be included. It is sometimes convenient to take G_0 as the smallest
subgroup that will generate the tiling.

Note 3. Reverse isometries sometimes appear, as in the triangulations
described below, in which each triangle is reflected in its sides.

Note 4. A simple arrangement that violates the second proviso is obtained by starting with a rectangular tiling and then drawing a diagonal in
each rectangle, but choosing the orientations of those diagonals randomly
from one rectangle to the next, as in Fig. 3.11b. Then all the triangles are
congruent, but they cannot be obtained by means of a group G_0 as described above.

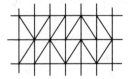

Fig. 3.11b.

Tilings of the sphere or of the hyperbolic plane are similar to tilings of the Euclidean plane, except that $\mathbb{G}$ is then the group of isometries of the sphere or of the hyperbolic plane, and the polygons have an angular excess in the spherical case and an angular defect in the hyperbolic case. Tilings of the sphere can be obtained by enclosing the sphere in any of the classical regular solids and projecting the faces and edges of that solid onto the sphere by radial lines drawn to the center of the sphere.

We now consider tilings by triangles; they will be called *triangulations* (of the sphere or plane). We further assume that the triangles can be obtained from each other by reflections in lines along the sides of the triangles; that gives a *reflection-triangulation*. We start with a given point P_0 and a given triangle having P_0 as a vertex. By successive reflections, we get a set of triangles as shown in Fig. 3.11c, and so on. Unless the triangles are isosceles (which we do not assume), there will be an even number of triangles meeting at P_0. The same is true of the other vertices, and it follows that the angles of the triangle have to be of the form $\pi/\ell, \pi/m, \pi/n$, where ℓ, m, n are integers greater than or equal to 2.

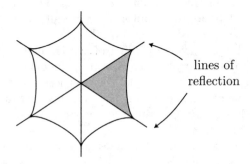

lines of
reflection

Fig. 3.11c.

The group G_0 of isometries is generated by the reflections referred to above. (We recall that, on the sphere, two reflections give a rotation;

on the Euclidean plane they can give a rotation or a translation, and on the hyperbolic plane they can give a rotation, a translation, or an ideal rotation.)

For the Euclidean case, the sum of the angles is equal to π; hence we must have

$$\frac{1}{\ell} + \frac{1}{m} + \frac{1}{n} = 1;$$

it is easily seen that there are just three possibilities

$$
\begin{aligned}
&\frac{1}{3} + \frac{1}{3} + \frac{1}{3} && (60° - 60° - 60°) \\
&\frac{1}{2} + \frac{1}{4} + \frac{1}{4} && (90° - 45° - 45°) \\
&\frac{1}{2} + \frac{1}{3} + \frac{1}{6} && (90° - 60° - 30°)
\end{aligned}
\qquad (3.11\text{-}1)
$$

The tiling that corresponds to the last of these is shown in Fig. 3.11d.

In the spherical case, we have

$$\frac{1}{\ell} + \frac{1}{m} + \frac{1}{n} > 1, \qquad (3.11\text{-}2)$$

and the possibilities are

$$
\begin{aligned}
&\frac{1}{2} + \frac{1}{2} + \frac{1}{n} && (n \geq 2), \\
&\frac{1}{2} + \frac{1}{3} + \frac{1}{n} && (n = 3, 4, 5).
\end{aligned}
$$

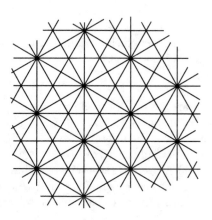

Fig. 3.11d.

Except for the first of these cases, when $n \geq 3$, each of these triangulations can be obtained from one of the classical regular solids by suitably triangulating the faces of that solid. For example, the regular dodecahedron has 12 faces, each of which is a regular pentagon. Each pentagon can be divided into ten triangles, so that the sphere is triangulated into 120 triangles, all having angles $\pi/2, \pi/3$, and $\pi/5$. The case $1/2 + 1/2 + 1/n$, with $n \geq 3$ can be visualized by dividing the equator of the sphere into $2n$ equal parts and then connecting the points of subdivision to the poles by arcs of great circles, as in Fig. 3.11e.

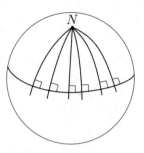

Fig. 3.11e.

In the hyperbolic case, we have

$$\frac{1}{\ell} + \frac{1}{m} + \frac{1}{n} < 1, \tag{3.11-3}$$

and now there are infinitely many possibilities for each of the integers ℓ, m, n. As indicated in Exercise 1 below, when positive angles α, β, γ are given such that $\alpha + \beta + \gamma < \pi$, there is always a triangle in the hyperbolic plane with those angles. In fact, such a triangle can be obtained by a simple construction. Each of the choices (3.11-3) leads to a triangulation, and we state that fact as a theorem.

Theorem 3.16: If we are given a triangle Δ_0 with angles $\pi/\ell, \pi/m, \pi/n$, where the integers ℓ, m, n satisfy (3.11-3), if we obtain further triangles by reflection in the sides of Δ_0, then obtain further triangles by reflections in the sides of the new ones, and so on, the result is a set of infinitely many triangles, no two of which overlap (adjacent ones have common boundary points), and which together cover the entire hyperbolic plane. The entire structure is invariant with respect to each of the reflections mentioned.

We shall give the proof only in the relatively simple case of equilateral triangles with 45° angles. The proof for the general case can be found in Caratheodory, *Theory of Functions of a Complex Variable, Vol. II*, Chelsea,

New York, 1964. It is longer than the proof given here because of the large number of special cases that have to be considered, but it follows the same principles.

The triangles to be considered have angles of $45° = \pi/4$, hence, according to Chapter 6, have sidelength equal to $1.5288570918\cdots$ (we shall not use this number).

We choose an arbitrary point O in the hyperbolic plane as origin, and we construct the eight triangles with O as a vertex in such a way that each is a reflection of its neighbors, as indicated in Fig. 3.11f. These are called triangles of the first order. Their outline is an octagon with 90° angles and is denoted by $\mathcal{P}_1$. About each of the vertices of $\mathcal{P}_1$ we construct similarly six more triangles with centers at that vertex, as in Fig. 3.11g. One of the six is shared with the next vertex (say clockwise) of $\mathcal{P}_1$, hence there are 40 new triangles, which are called triangles of the second order.

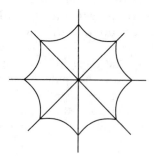

Fig. 3.11f.

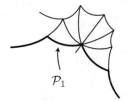

Fig. 3.11g.

The outline of these triangles of the second order is a polygon $\mathcal{P}_2$ of 32 vertices, at 8 of which the internal angle is 135° and at 24 of which it is 90°. We continue this procedure, to construct triangles of order $3, 4, \ldots, k, \ldots$, and polygons $\mathcal{P}_3, \mathcal{P}_4, \ldots, \mathcal{P}_k, \ldots$. Clearly each of the polygons lies inside the next, the origin O is at the center of all of them (the entire structure is invariant with respect to eightfold rotations about the origin O), and the triangles of order k lie between $\mathcal{P}_{k-1}$ and $\mathcal{P}_k$. Each vertex of $\mathcal{P}_k$ is a vertex of either two or three of the triangles of order k; hence the interior angles of the polygons are all either $90°$ or $135°$. It follows just as in the Euclidean case that the region bounded by each of these polygons is convex, since all the angles are less than $180°$. Clearly, the triangles of order k fill up the region between $\mathcal{P}_{k-1}$ and $\mathcal{P}_k$, and we now prove by an induction on k what may seem obvious on intuitive grounds, but really has to be proved:

No triangle of order k overlaps any other triangle of order k, by which we mean, of course, that no two of them have any interior points in common.

To prove this, we note that it is obviously true for triangles of the first order; we assume that it is true for triangles of order $\leq k-1$, and we prove that it is then true for the triangles of order k. We consider first those triangles of order k that have a side in common with the polygon $\mathcal{P}_{k-1}$, and we call them $Z_1, Z_2, \ldots$; we prove that they do not overlap each other. Let $V_0, V_1, \ldots, V_q = V_0$ be the vertices of $\mathcal{P}_{k-1}$, taken, say, in clockwise order. It is shown in Exercise 2 below that the angular regions $\angle V_{j-1}OV_j$ $(j = 1, \ldots)$ are nonoverlapping, and we show that precisely one of the triangles of type Z lies in each of them. Let Z be the triangle with segment $V_{j-1}V_j$ as a side. The angle β shown in Fig. 3.11h is $\leq 3 \cdot 45°$, and we see that the triangle Z is to the right of the ray OV_{j-1}.

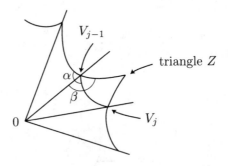

Fig. 3.11h.

Similarly, it is to the left of the ray OV_j, hence it lies in the angular region $\angle V_{j-1}OV_j$. Therefore, the triangles of type Z do not overlap. Now let Z_{j-1} and Z_j be two adjacent triangles of type Z, with outer vertices W_{j-1} and W_j and let $E_1, E_2, \ldots$ be the other triangles of order k (either 3 or 4 in number) that have a vertex at the vertex V_j of $\mathcal{P}_{k-1}$, as shown in Fig. 3.11i. Since triangle Z_{j-1} lies in $\angle V_{j-1}OV_j$, the ray OW_{j-1}, passes through Z_{j-1}; hence the angle $\angle OW_{j-1}Q$ is less than $90°$, and we see that the triangles $E_1, E_2, \ldots$ all lie in the angular region $\angle W_{j-1}OW_j 1/M$ so they do not

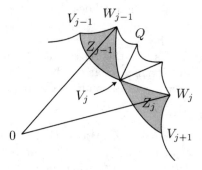

Fig. 3.11i.

overlap any triangles E_1', E_2' that have vertices at any other of the points V_i. By construction, the triangles $E_1, E_2, \ldots$ that have a vertex at V_j do not overlap each other and do not overlap Z_{j-1} or Z_j and we see that the triangles or order k are all nonoverlapping, as required.

It remains to show that the region R covered by all the equilateral triangles generated in this way is the entire hyperbolic plane. To that end, we show that R has no boundary. Let P be *any* point of R. Then P is at a distance $\leq a$ from a vertex of one of the triangles, where a is the sidelength $(= 1.528570918\cdots)$ but that vertex is surrounded by eight of the triangles, so P is an *interior* point of R. That is, R has no boundary points and hence is the entire plane. The invariance of the structure under reflections can be seen in the following way: If the common edge of any two triangles is continued in either direction as a straight line, it continues to lie along common edges of triangles that are reflections of each other as in Fig. 3.11j.

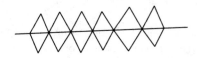

Fig. 3.11j.

From the laws of congruence, it follows that the entire construction on one side of that line is identical with the reflection of the construction on the other side.

From the theorem and the earlier comments on the more familiar Euclidean case, we arrive at the following conclusion:

Theorem 3.17 (*Contrast Between the Euclidean and the Hyperbolic Planes*):

In reflection triangulations of the Euclidean plane, there are just three possible shapes of the triangles, but they can be of arbitrarily small size (or of arbitrarily large size). In the hyperbolic case, there are infinitely many possible shapes of the triangles, but their sizes are determined by their angles, and the smallest possible triangle that can be used has an angular defect (which is equal to its area in suitable units according to Section 3.12) equal to $\pi/42$.

Closely related to tilings are lattices. Given a point P_0 and a subgroup G_0 of $\mathbb{G}$, the set $\mathcal{L} = \{P_0, P_1, \dots\}$ of the images of P_0 under all the isometries of G_0 is a *lattice* if the following conditions are satisfied:

(1) The point set $\mathcal{L}$ is discrete, which means that for each point P_j of $\mathcal{L}$, there is a distance $\varepsilon > 0$ so that no other point of $\mathcal{L}$ is within a distance ε of P_j.
(2) For each point P_j there is a polygon $\mathcal{P}$ whose vertices are points of $\mathcal{L}$ and which contains P_j in its interior.

Note 1. $\mathcal{L}$ can also be described as the images of *any* given point P_j of $\mathcal{L}$, not necessarily the originally given point P_0.

Note 2. To establish the discreteness, it suffices to show that condition (1) holds for any single point of $\mathcal{L}$, for example for P_0.

Note 3. If the point P_0 is moved to a point P_0' not in $\mathcal{L}$, a different lattice $\mathcal{L}'$ is obtained, for the given group. For example, if G_0 is the group associated with the Euclidean tiling shown in Fig. 3.11d, lattices with quite different appearances are obtained by taking P_0 as (1) a six-fold center of rotation, (2) a three-fold center of rotation, (3) a two-fold center of rotation, or (4) a general point in the interior of one of the triangles. *Hint:* Sketch these lattices.

Note 4. Requirement (2) serves to exclude cases in which $\mathcal{L}$ consists merely of equally spaced points on a single line or equally spaced points on a single circle.

The group G_0 associated with any tiling determines a lattice (in general, many lattices); the converse is true only if we include cases in which the regions $R_0, R_1, \dots$ are unbounded. A triangle (or polygon) one or more of whose vertices is an ideal point at infinity is called *degenerate*. Consider a triply degenerate triangle obtained by taking any three distinct ideal points, say at angles $\theta_1, \theta_2, \theta_3$ in some polar coordinate system, and joining those ideal points in pairs by straight lines, according to Theorem 3.13 in Section 3.4. One's first thought is that one can get triangles of different "shapes" by different choices of θ_1, θ_2 and θ_3. However, one can then choose a new coordinate system r', θ' with an origin at the center of the triangle, as described in Exercise 11 below, then the triangle has a threefold symmetry of rotation about the new origin, and the angles θ_1', θ_2', and θ_3' differ in pairs by 120°. A tiling by such triangles is the subject of Exercise 12. Degenerate triangles that satisfy (3.11-3) correspond to taking at least one of the integers ℓ, m, n as infinity and at least one of the others greater than 2.

Exercises

1. Show that if α, β, γ are positive and $\alpha + \beta + \gamma < \pi$, there exists in the hyperbolic plane a triangle with angles α, β, γ. *Hint:* Do it first for the case $\alpha = 90°$; on one of two orthogonal lines construct a ray $\overrightarrow{k}$ at distance y and at angle β, as shown in Fig. 3.11k. If y is small enough, $\beta < \Pi(y)$, so that the ray intersects the vertical line. From the properties of the triangle functions described in Section 2.5, show that for some (unique) value of y, the desired triangle is obtained.

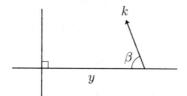

Fig. 3.11k.

2. Show that if rays $\overrightarrow{k}, \overrightarrow{k'}, \overrightarrow{k''}$ emanate from a point P, with $\overrightarrow{k'}$ between $\overrightarrow{k}$ and $\overrightarrow{k''}$, then the regions interior to the angles $\angle \overrightarrow{k}, \overrightarrow{k'}$ and $\angle \overrightarrow{k'}, \overrightarrow{k''}$ do not overlap.
3. Consider how to generalize the argument in the text to a triangulation by equilateral triangles with angles π/n, for $n > 4$.
4. Show how to choose ℓ, m, n so that the angular defect of the triangle is $\pi/42$, and show that the angular defect can never be less than this, for integer ℓ, m, n.
5. Let Δ_0 be the triangle with angles $\pi/\ell, \pi/m, \pi/n$ at vertices A, B, C. Show that the subgroup G_0 of $\mathbb{G}$ generated by reflections in the sides of that triangle contains rotations R_A, R_B, R_C through angles $2\pi/\ell, 2\pi/m, 2\pi/n$ about those vertices.
6. In the notation of the preceding exercise, show that if a subgroup of $\mathbb{G}$ contains the rotations R_A and R_B, then it automatically contains R_C.
7. Show that, although the subgroup generated by the rotations R_A, R_B, R_C cannot be all of the subgroup G_0 generated by reflections in the sides of the triangle Δ_0, the subgroup generated by R_A and R_B together with reflection in the line AB is all of G_0.
8. Let R_A and R_B be the rotations in the preceding exercises, but now let the distance $y = |AB|$ be variable. For what values of y is the subgroup generated by R_A and R_B the subgroup of a lattice? *Hint:* Unless y is properly chosen, the point set consisting of all images of a fixed point P_0 may not be discrete.

9. Prove or disprove: If the subgroup G_0 connected with a tiling contains only direct isometries, then G_0 can be generated by two suitably chosen isometries. What about the case where G_0 also contains reverse isometries (i.e., ones containing reflections)?

10. Let G_0 be a subgroup of $\mathbb{G}$ associated with a tiling of the plane (Euclidean or hyperbolic), and let S be the set of all images of a given point P_0 under the isometries of G_0. Show that there are only finitely many points of $\mathcal{L}$ in each of the regions R_j (hence the set is discrete, therefore a lattice). Under what conditions does each R_j contain exactly one point of S?

11. Let Δ_0 be a triply degenerate triangle with vertices at infinity and with sides ℓ_1, ℓ_2, ℓ_3. The sides ℓ_2 and ℓ_3 are asymptotic at one of the vertices and orthogonal to a horocycle with that ideal point as center. Let P be the midpoint of the horocycle segment between ℓ_2 and ℓ_3, and let m_1 be the line through P asymptotic to ℓ_2 and ℓ_3. We may regard m_1 as the angle bisector of the sides ℓ_2 and ℓ_3. Let m_2 be similarly the angle bisector for sides ℓ_3 and ℓ_1. Show that the intersection Z of m_1 and m_2 is equidistant from ℓ_1 and ℓ_2, so that the third angle bisector also passes through Z. Show by consideration of congruent triangles that Z is equidistant from the three sides of the triangle and that the triangle has threefold symmetry of rotation about Z, so that, in terms of polar coordinates with origin at Z, the angles of the ideal vertices are separated pairwise by $120°$.

12. Let Δ_0 be the triply degenerate triangle of the preceding exercise. Consider the set of triangles obtained by reflecting Δ_0 in its sides, then reflecting the new triangles in their sides, and so on. Let $P_0, P_1, \ldots,$ be the centers of those triangles. Show that those points form a lattice.

13. Consider similar tilings by singly or doubly degenerate triangles.

14. Show that there is no lattice associated with any subgroup of the 2-parameter group of isometries discussed in Section 3.9. If the subgroup is large enough so that condition (2) for a lattice is satisfied, that is, so that the lattice doesn't reduce to a set of points on a single line or a single curve, then condition (1) fails. The set of points is not discrete.

3.12 Area and Angular Defect; Equivalence of Polygons

In both Euclidean and hyperbolic geometry (also spherical), the intuitive concept of area is subject to the following fundamental principles:

(1) It is additive; if two regions are placed side by side so as to make a larger region, then the area of the larger region is the sum of the areas of the two smaller ones.

(2) Congruent figures have the same area.

From these, it follows, for example, that if a polygon is triangulated, that is, is represented as the union of several nonoverlapping triangles as in Theorem 2.15 of Section 2.9, and if those triangles are then moved around in the plane, without change of size or shape, and then reassembled in a different arrangement to make a second polygon, then the two polygons, which are said to be *equivalent by triangulation*, have the same area.

In Euclidean geometry, formulas are derived for the areas of squares, rectangles, triangles, and so on. One starts with squares, as indicated by the expressions "square inch," "square meter," etc. One observes that a square of sidelength L is exactly the union of n^2 squares of sidelength L/n; hence one defines the area of any square as equal to the square of its sidelength. The given square has been covered by a *tiling* by smaller squares (tilings in general were discussed in the preceding section). Then a rectangle whose length and width have a rational ratio can be represented as the union of a number of smaller squares, from which one decides (via a limiting process, in which the dimensions of the small squares tend to zero, in order to cover the case of an irrational ratio) that the area of a rectangle is equal to the product of its two dimensions. Then, given a triangle, one reduces the problem to that of a rectangle in two steps: first, one doubles the area by adjoining another triangle obtained by rotating the first one through 180° about the midpoint of one of its sides, to form a parallelogram. Any parallelogram can be made into a rectangle by adding and subtracting congruent triangles. In this way one decides that the area of a triangle is one-half its base times its altitude.

The discussion in the hyperbolic case is similar, but with the difference that there are no squares in the hyperbolic plane, and, more generally, no polygons that can be decomposed into smaller identical polygons similar to the given one. In fact, there are no similar polygons except congruent ones. As pointed out in Section 3.11, there are no tilings of the hyperbolic plane by arbitrarily small "tiles." The smallest possible triangle that can be used for a certain kind of triangular tiling has angular defect equal to $\pi/42$; smaller ones don't fit properly.

In the hyperbolic case, we show first that if two polygons are equivalent to a third by triangulation, then they are equivalent to each other. (That holds also in the Euclidean case.) Then, we show that if two triangles have the same angular defect and a common base, they are both equivalent to a certain quadrilateral, and hence are equivalent to each other. Then, we show that the requirement of a common base can be omitted so that any two triangles with the same defect are equivalent, and hence have the same area. It then follows, since area and defect are additive, that the area of a triangle, or of any polygon, is proportional to its angular defect. Finally, we establish the proportionality constant by considering the limit of very small triangles and requiring that in that limit the area be equal to one-half the base times the altitude (Exercise 5 in Section 6.3). Then, if the unit of length is so chosen that $K = 1$, the area is *equal* to the defect. Hence, the area of a triangle cannot exceed π and the area of a triangle with its vertices at infinity (at ideal points) is equal to π.

Theorem 3.18: If each of two polygons is equivalent to a third, then the two are equivalent to each other.

Sketch of the Proof. Since any polygon can be triangulated, the relation of equivalence, as we have defined it, is reflexive. It is also evident from the definition that the relation is symmetric: if $\mathcal{P}$ is equivalent to $\mathcal{Q}$, then $\mathcal{Q}$ is equivalent to $\mathcal{P}$. What we have to prove is that the relation is transitive: if $\mathcal{P}$ is equivalent to $\mathcal{Q}$ and $\mathcal{Q}$ is equivalent to $\mathcal{R}$, then $\mathcal{P}$ is equivalent to $\mathcal{R}$. In the two given equivalences, the intermediate polygon $\mathcal{Q}$ gets in general two triangulations. If these triangulations are superposed by drawing in $\mathcal{Q}$ simultaneously all the sides and vertices of both triangulations, then $\mathcal{Q}$ is represented as made up of many small figures, some of which may be triangles and other more general polygons. Then, we triangulate the more general polygons. The result is many tiny triangles, which, when put together in one order, give a representation of $\mathcal{P}$ and when put together in another order, give a representation of $\mathcal{R}$. Hence $\mathcal{P}$ and $\mathcal{R}$ are equivalent.

Theorem 3.19: Let Δ be any triangle. Let a be the length of one of its sides. Then, Δ is equivalent to a unique Saccheri quadrilateral with sidelength a between its acute angles and with defect equal to the defect of Δ.

Sketch of the Proof. Let ℓ be the line joining the midpoints of the other two sides, as shown in Fig. 3.12a. Drop perpendiculars (dashed lines) from the vertices of the triangle to that line. It is then left to the reader to show that the smaller right triangles in Fig. 3.12a are congruent in pairs. That proves the equivalence (hence the equality of the defects) and shows that the quadrilateral $BB'CC'$ is a Saccheri quadrilateral, that is, its angles at B' and C' are right angles and $|BB'| = |CC'|$, so that its angles at B and C are equal acute angles. If the vertex A is moved along an equidistant to the line ℓ, keeping the distance $|AA'|$ fixed, all the resulting triangles are equivalent to the same Saccheri quadrilateral. That proves the following:

Corollary: Two triangles with a common side length are equivalent if and only if they have the same angular defect.

We wish to show now that two triangles with the same defect are equivalent even if they don't have a common side length. Let Δ and Δ' be the triangles, with a, b, c the side lengths of the first and a', b', c' the side lengths of the second. We shall show how to construct a triangle Δ'' with the same angular defect as the given triangles and with one side of length a and one of length b'. According to the Corollary, Δ'' is equivalent to each of the given triangles; hence they are equivalent to each other, by

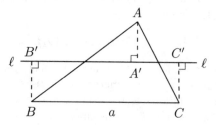

Fig. 3.12a.

Theorem 3.18. Let Δ be the triangle ABC of Fig. 3.12a, with $a = |BC|$ and $b = |AC|$. We assume that $b < b'$ (otherwise, we interchange the two triangles; if $b = b'$, the two triangles are equivalent by the Corollary and there is nothing further to prove). In Fig. 3.12a, we draw a circle about vertex C with radius $1/2\ b'$, which is greater than $1/2\ b$, and hence greater than $|CC'|$, so that the circle intersects the line ℓ in two points. Let X be one of the points and extend CX an equal distance to point Y, as shown in Fig. 3.12b. By consideration of congruent triangles, it is easy to see that Y is at the same distance from the line ℓ as the point A in Fig. 3.12a. By the proof of Theorem 3.19, it is seen that BCY has the same defect as the given triangles, and its side YC is of length b'; hence it is the required triangle Δ'', and we have proved the following:

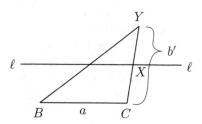

Fig. 3.12b.

Theorem 3.20: Two triangles are equivalent if and only if they have the same defect.

It was noted above that, since the equivalence of polygons involves triangulation, equivalent polygons have equal area, under any definition of area which is additive and is such that congruent triangles have equal area. It then follows from the above theorem that, at least for triangles, the area

A depends only on the angular defect δ; $A = A(\delta)$. From the additivity of both area and defect, we have

$$A(\delta_1 + \delta_2) = A(\delta_1) + A(\delta_2).$$

It follows, since area and defect are inherently positive quantities for any triangle, that the function $A(\delta)$ is increasing. It is a theorem of calculus that any such function [additive, increasing, and defined for all δ in an interval such as $(0, \pi)$] is homogeneous linear; that is, $A(\delta)$ is equal to a constant times δ. It then follows further, from the additivity of area and angular defect, that this result holds not only for triangles, but also for other polygons, since any polygon can be triangulated. Hence, we have the following:

Corollary: As in the Euclidean case, area is uniquely defined, up to a multiplicative constant, by the requirements that it be additive and that congruent figures have equal areas.

As a final comment, it can be proved that any two polygons (not necessarily triangles) with the same angular defect are equivalent. That is known as Bolyai's Theorem; see *The Foundations of Geometry and the Non-Euclidean Plane*, by George Martin, Chapter 33, (Springer Verlag, Berlin, 1975). That is of course a statement about the existence of triangulations of the two polygons such that the conditions in the definition of equivalence are satisfied. An alternative statement is that any two polygons with the same area are equivalent.

An alternative approach is described in *Non-Euclidean Geometry*, by H.S.M. Coxeter (University of Toronto Press, Fourth Edition, 1961), Chapter XIII, based on certain unbounded figures with finite area. It is shown that a horocycle sector (a figure bounded by an arc of a horocycle and two asymptotic lines) is equivalent to a certain bounded region.

The conclusions reached in this section are also obtained in Chapter 5 by the methods of differential geometry.

3.13 A Misunderstanding About Astronomical Parallax

In some books there is an argument that appears to go as follows: Let C be a distant star, and let B be a point of the earth's orbit such that the angle $\angle BAC$ is a right angle, where A is the center of the earth's orbit. (See Fig. 3.13. Can you prove that there always *is* such a point B?) It is then said that the astronomers measure the angle $\angle ABC$, whose complement is the parallax of the star. Then the angular defect of the triangle ABC would be less than that parallax. For a few nearby stars, the parallaxes have been measured and are found to be a few hundredths of a second of arc; hence, it is concluded that there exist very large triangles with defect less than a few hundredths of a second. Such an analysis is misleading, for the following reasons:

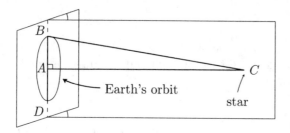

Fig. 3.13.

(1) The astronomers don't measure the angle $\angle ABC$. There is no way of "sighting along" the line BD, which is the diameter of the earth's orbit. (It takes the earth six months to get from D to B.) The astronomers could in principle calculate the angle $\angle ABC$ to a certain accuracy (*not* hundredths of a second) by a complicated calculation which would take into account the motion of the earth in its orbit, the perturbation of that orbit by the moon and Jupiter, together with observations of the stars and planets. Surely, the angle would turn out to be 90° to within the accuracy of the calculation.

(2) If the astronomers could calculate $\angle ABC$ for some star and if we were to use the result to put an upper bound on the defect of a triangle, we would not use for that purpose a nearby star, such as the ones for which parallax has been measured, but a more distant star.

(3) Astronomical parallax is a purely *relative* measurement made by taking two photographs of a given star field six months apart. Microscopic examination of the photographs reveals that all the stars are in exactly the same positions *with respect to each other*, except for a few bright stars, which have been displaced slightly, in the six months, relative to the rest. The displacement corresponds to a few hundredths of a second of arc, as seen from the center of the earth's orbit. That displacement is then interpreted, *according to Euclidean geometry*, to determine the distances to those nearby stars.

(4) The conceptually relevant measurement of angular defect is the one allegedly made by Gauss (though never published) on a triangle whose vertices were at the summits of three mountains in Germany. The defect was found to be zero to the accuracy of the measurement. Although that triangle was small and the accuracy was low, the interpretation was clear and correct. It is remarkable that Gauss had the imagination

and the audacity needed to make that measurement, 100 years before Einstein.

(5) Quite a bit is known about the geometry of the space-time in which we live, from the general theory of relativity; it is neither Euclidean nor hyperbolic.

Exercise

1. To get an idea of the magnitudes involved, calculate how far away an apple would have to be to subtend an angle of one-hundredth of a second of arc.

3.14 Bounds on Angular Defects of Small Triangles and Quadrilaterals

We have seen that under any reasonable definition of area, the angular defect of a triangle is proportional to its area. That suggests, by analogy with Euclidean geometry that as triangles get smaller and smaller the defects should decrease as the square of the linear dimensions. We have in fact the following:

Theorem 3.21: There is a constant C_1 such that if L is the length of the longest side of a triangle, then the angular defect is less than $C_1 L^2$.

Theorem 3.22 (Thin Triangles)**:** There is a constant C_2 such that if L is the length of the longest side of a triangle and is less than 1, and α is its smallest angle, then the defect is less than $C_2 L^2 \alpha$.

Remark: Our method of proof uses Lambert quadrilaterals. We recall from the Remark in Section 3.5 that if $ABCD$ is such a quadrilateral with right angles at vertices A, B, C and with side lengths a, a', b, b' as in Fig. 3.14a, then (1) the angle at D is acute, and (2) $a < a'$ and $b < b'$.

We denote by $\delta(a, b)$ the angular defect of a Lambert quadrilateral with "altitude" a and "base" b (these are distances between pairs of *right* angles). Clearly, $\delta(a, b) = \delta(b, a)$, because $ABCD$ is congruent to $CBAD$ under a reverse isometry that interchanges "base" and "altitude," for example under a reflection in the line that bisects the angle at B. In the proofs of the theorems, we shall make repeated use of the following lemma:

Lemma 3.3: $\delta(1/2\, a, b) < 1/2\, \delta(a, b)$ and $\delta(a, 1/2\, b) < 1/2\, \delta(a, b)$.

Proof: Let E be the midpoint of the altitude, and erect the perpendicular EF as shown in Fig. 3.14b. According to the Remark above, EF is longer than BC. Mark off a distance $|EG|$ equal to $|BC|$ on EF, and erect a

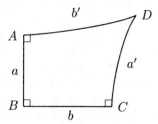

Fig. 3.14a.

perpendicular to EF at G, as shown. Then, $AEGH$ is congruent to $EBCF$. According to Section 3.2, if one polygon lies inside another, it has a smaller defect; hence, the defect of $EBCF$ is less than half the defect of $ABCD$. That proves the first statement in the Lemma, the proof of the second is similar.   □

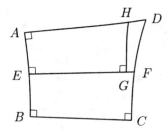

Fig. 3.14b.

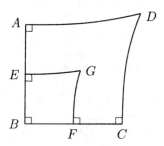

Fig. 3.14c.

Proof of Theorem 3.21. Let $\mathcal{Q}_0 = ABCD$ be an isosceles Lambert quadrilateral (i.e., one with $a = b$), and call $a_0 = a = b$. Let E and F be the midpoints of the sides AB and BC, erect perpendiculars to those sides at those points as shown in Fig. 3.14c, and let G be the intersection of those two perpendiculars. Call $\mathcal{Q}_1 = EBFG$, which is also an isosceles Lambert quadrilateral. By two applications of Lemma 3, we see that the defect of $\mathcal{Q}_1$ is less than one-fourth the defect of $\mathcal{Q}_0$. We then divide $\mathcal{Q}_1$ into four still smaller quadrilaterals in a similar manner, and call the lower left of those $\mathcal{Q}_2$, and so on. If δ_n is the defect of Q_n and a_n is its side length (base or altitude), then $a_n = 2^{-n}a_0$ and $\delta_n < 4^{-n}\delta_0$, hence $\delta_n < \delta_0(a_n/a_0)^2$. Now let Δ be any triangle with longest side of length L. First, assume that $L < a_0$ and choose n so that $a_{n+1} \leq L < a_n$. Then, Δ can fit inside $\mathcal{Q}_n$; hence its defect is less than $\delta_n < \delta_0 a_n^2/a_0^2 = 4\delta_0 a_{n+1}^2/a_0^2 \leq C_1 L^2$, where $C_1 = 4\delta_0/a_0^2$. Lastly, if $L \geq a_0$, then, since the defect of a triangle can never exceed π, we can ensure that the relation defect is less than $C_1 L^2$ continues to hold, for all L, by increasing the value of C_1 if necessary. This completes the proof of Theorem 3.21.

Proof of Theorem 3.22: We denote by Δ_0 an isosceles right triangle with legs of length $= 1$, hence hypotenuse < 2, and acute angle $\alpha_0 < \pi/4$, angular defect $\delta_0 = \pi/2 - 2\alpha_0 < C_1(2)^2 = 4C_1$, by Theorem 3.21. We then denote by Δ_n an isosceles right triangle of leg length 2^{-n}, defect $\delta_n < 4^{-n+1}C_1$ and acute angle $\alpha_n > \alpha_0$. We bisect the acute angle of this triangle and call $\Delta_{n,1}$ that one of the resulting smaller triangles with smaller defect; then we bisect the angle of that triangle and call $\Delta_{n,2}$ that one of the still smaller triangles with smaller defect, and so on. Then, $\Delta_{n,k}$ has defect $\Delta_{n,k} < 2^{-k}4^{-n+1}C_1$ and two sides both of length greater than or equal to 2^{-n}, and smallest angle is $2^{-k}\alpha_n > 2^{-k}\alpha_0$.

We now let Δ be a given triangle with maximum side length $L < 1$ and smallest angle $\alpha < \alpha_0$. We choose n such that $2^{-n-1} \leq L < 2^{-n}$, and we choose k such that $2^{-k-1}\alpha_0 \leq \alpha < 2^{-k}\alpha_0$. Then Δ can fit inside $\Delta_{n,k}$, and hence its defect is

$$\delta < 2^{-k}4^{-n+1}C_1 = 2 \cdot 2^{-k-1}16 \cdot 4^{-n-1}C_1 < 32\frac{\alpha}{\alpha_0} \; L^2 C_1 = C_2 L^2 \alpha,$$

as required. Lastly, to cover cases in which $\alpha \geq \alpha_0$, we increase the value of the constant C_2, if necessary. Only a finite increase is required, because the defect δ is in no case greater than π. □

We note in passing that it is not so easy to allow for values of L greater than 1, because, as $L \to \infty$, $\alpha_0 \to 0$. In applications, we are only interested in cases in which L and α are both small.

By a slight generalization of the procedure in the proof of Theorem 3.21, we have the following result:

Theorem 3.23: For the defect of a Lambert quadrilateral of altitude a and base b, we have

$$\delta(a,b) < C_1 ab.$$

Proof: We start, as before, with an isosceles Lambert quadrilateral of altitude and base equal to a_0; we now call it $\mathcal{Q}_{0,0}$. We divide its altitude into 2^n parts and its base into 2^k parts, and we denote the resulting lower left smaller quadrilateral by $\mathcal{S}_{n,k}$; it has altitude $2^{-n}a_0$ and base $2^{-k}a_0$. A given Lambert quadrilateral with altitude a and base b can fit into $\mathcal{Q}_{n,k}$ if $2^{-n-1}a_0 \le a < 2^{-n}a_0$ and $2^{-k-1}a_0 \le b < 2^{-k}a_0$; hence $\delta(a,b) < 2^{-n-k}\delta_0$. From here on, the argument is the same as in Theorem 3.21, and we have $\delta(a,b) < C_1 ab$, as required. $\square$

3.15 Length of a Circular Arc and Area (Defect) of a Circular Sector

In Sect. 3.9, a procedure was described for determining the length of a curve. [See equation (3.9-2).] Here we shall apply that procedure to circular arcs. The portion of a circle in the interior of an angle with vertex at the center of the circle is a *circular arc*. We wish to find an expression for its length in terms of the angle α and the radius r.

A closely related quantity is the area of the sector bounded by the circular arc and the two radii shown in Fig. 3.15a, interpreted as the limit of the area of a polygon consisting of the two radii and a polygonal approximation to the arc as the number of points on the arc increases and their separation tends to zero. Since we have shown that the area of a polygon is proportional to its angular defect, we shall take the limit of the defect of the polygon and call it the *defect* of the circular sector; it is the area of the sector in suitable units.

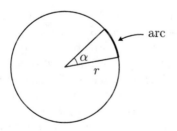

Fig. 3.15a.

We shall find that, for a given radius r, the length of the arc and the defect of the sector are both proportional to the angle α. That is, there are functions $f(r)$ and $g(r)$ (formulas not yet known) such that

$$\text{defect of the sector} = f(r)\alpha,$$
$$\text{length of the arc} = g(r)\alpha.$$

[In Euclidean geometry, we have $g(r) = r$ and $f(r) = r^2/2$, for the usual unit of area. The corresponding formulas for $f(r)$ and $g(r)$ in the hyperbolic case are derived in Chapter 6.]

We first settle a question that will arise in the course of the discussion. In the hyperbolic plane, it is possible for two tangents to a circle to fail to meet, even if the angular separation of the points of tangency is less than 180° as indicated in Fig. 3.15b. We show, however, that if that angular separation is not greater than a certain angle (which generally depends on the radius), then the tangents *do* meet.

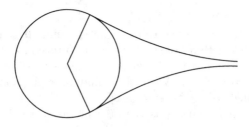

Fig. 3.15b.

Lemma 3.4: For a circle of radius r there is an angle $\omega = \omega(r)$ such that if two points of the circle are separated in angle by not more than ω, then the tangents meet.

Proof. Let ZA be a radius, where Z is the center of the circle, and let ℓ be the line through A perpendicular to the radius as in Fig. 3.15c. It follows from Theorem 2.2 that ℓ has only one point in contact with the circle, that is, it is a tangent. Let P be a point on ℓ at distance $|AP| = |ZA| = r$ from A, so that the triangle ZAP is isosceles. We denote the acute angles of that triangle by $\omega/2$ (they are less than 45°). We then draw another radius ZB at the same angle $\omega/2$ with respect to ZP, but on the other side, and we join the points B and P, as shown. It is seen from the SAS criterion that the two triangles shown are congruent, so that BP is also a tangent. It is thus seen that if points of tangency are separated by angle ω the tangents meet (in fact, at the point P) and it is easy to show that if the points of

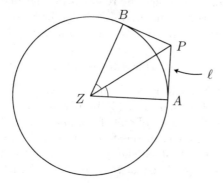

Fig. 3.15c.

tangency are separated by any smaller angle, the tangents also meet, as
required.   □

We consider first the problem of circular arclength.

Following Section 3.9, we approximate the circular arc of radius r and
angle α shown in Fig. 3.15a by a polygonal curve C_0 with vertices on the
circle at angles θ_i from the horizontal, where

$$\mathcal{P} : 0 = \theta_0 < \theta_1 < \cdots < \theta_n = \alpha$$

is a partition of the interval $[0, \alpha]$, and the length of C_0 is denoted by $s(\mathcal{P})$.
In terms of the function $\phi(a, b, \gamma)$ introduced in connection with Axiom 6
and discussed in Section 2.5, the length of the chord between angles θ_{i-1}
and θ_i is $\phi(r, r, \theta_i - \theta_{i-1})$; hence the length of the polygonal approximation
C_0 is

$$s(\mathcal{P}) = \sum_{i=1}^{n} \phi(r, r, \theta_i - \theta_{i-1}). \tag{3.15-1}$$

As in Section 3.9, we now consider a sequence of partitions $\mathcal{P}_k$, where $\mathcal{P}_{k+1}$
contains all the points θ_i of $\mathcal{P}_k$ and more too. Then, the numbers $s(\mathcal{P}_k)$
increase with increasing k. We assume also that the largest interval $\theta_i - \theta_{i-1}$
tends to zero as $k \to \infty$. It is easy to see that the quantities $s(\mathcal{P}_k)$ are
bounded: it is only necessary to consider a partition $\mathcal{P}_k$ such that each
$\theta_i - \theta_{i-1}$ is not greater than the quantity $\omega(r)$ of Lemma 4, and then to
define an exterior polygon obtained by placing tangents to the circle at
each point θ_i. It follows from the triangle inequality that the length of that
polygon is greater than all the $s(\mathcal{P}_k)$. Since the $s(\mathcal{P}_k)$'s form an increasing
bounded sequence, they have a limit, and we therefore define the length of
the circular arc as the quantity

$$s_0(\alpha) = \lim_{k \to \infty} s(\mathcal{P}_k). \tag{3.15-2}$$

We now consider two adjacent circular arcs, the first one with θ in $[0, \alpha]$ and the second with θ in $[\alpha, \alpha + \beta]$. The corresponding polygonal approximations together approximate the two arcs put together, and we conclude that

$$s_0(\alpha + \beta) = s_0(\alpha) + s_0(\beta).$$

From this we draw two conclusions: (1) since $s_0(\alpha)$ and $s_0(\beta)$ are positive, $s_0(\theta)$ is an increasing function of θ; (2) by dividing an arc into N equal parts and considering the length of one of the parts, we see that $s_0(\alpha) \to 0$ as $\alpha \to 0$; hence $s_0(\theta)$ is a continuous function. It is a theorem of calculus that a continuous function that satisfies the above additive law is linear and homogeneous; that is, $s_0(\alpha)$ is equal to a constant times α. The constant may of course depend on the radius r, which has been held fixed until now; hence we call it $g(r)$, so that $s_0(\alpha) = g(r)\alpha$. Furthermore, it follows from Theorem 2.14 in Section 2.8 that $g(r)$ is a nondecreasing function of r. Lastly, by choosing a partition so that all increments are the same, say equal to $\Delta\theta$, with $\alpha = n\Delta\theta$, we see that $\phi(r, r, \Delta\theta)$ is less than $g(r)n\Delta\theta$, but approaches that value as $\Delta\theta \to 0$ and $n \to \infty$, and hence $\phi(r, r, \Delta\theta)/\Delta\theta \to g(r)$ as $\Delta\theta \to 0$. We have proved the following:

Theorem 3.24: There is a nondecreasing function $g(r)$ such that the length $s_0(\alpha)$ of a circular arc of radius r and angle α is given by

$$s_0(\alpha) = g(r)\alpha.$$

Furthermore,

$$\frac{\phi(r, r, \Delta\theta)}{\Delta\theta} \text{ is } < g(r) \text{ and } \to g(r) \text{ as } \Delta\theta \to 0. \qquad (3.15\text{-}3)$$

We note in passing that the nondecreasing character of $g(r)$ does not hold in the geometry on a sphere of radius a: the corresponding function increases for $0 < r < (\pi/2)a$ and decreases for $(\pi/2)a < r < \pi a$.

The treatment of the problem of the defect (area in suitable units) of a circular sector is almost identical, word-for-word. We denote by $\delta(a, b, \gamma)$ the angular defect of a triangle with two side lengths a and b and included angle γ. Then we have the following.

Theorem 3.25: There is a function $f(r)$ such that the defect of a circular sector of radius r and angle α is given by

$$\delta_0(\alpha) = f(r)\alpha.$$

Furthermore,

$$\frac{\delta(r, r, \Delta\theta)}{\Delta\theta} \text{ is } < f(r) \text{ and } \to f(r), \text{ as } \delta\theta \to 0.$$

3.16 Bounds for $g(r)$ and $f(r)$ for Small r

In Chapter 6, we shall see that, as $r \to 0$, $g(r)/r \to 1$, and $f(r)/r^2 \to 1/2$. Here we show the weaker results that $g(r) < (6/\pi)r$ and $f(r) < \text{const } r^2$, for small enough r. Those bounds will be needed later.

We consider an isosceles right triangle with legs of length r, with one acute angle at the center of a circle of radius r and with hypotenuse horizontal, as in Fig. 3.16.

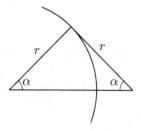

Fig. 3.16.

The angular defect is $\delta = (\pi/2) - 2\alpha$. According to Theorem 3.21 there is a constant L such that if $r < L$, then $\delta < \pi/6$, hence $\alpha > \pi/6$. Now, the length of the arc of radius r, angle α is clearly less than r, hence $g(r)\alpha < r$, $g(r) < (6/\pi)r$. The defect of the circular sector of radius r, angle α is, by Theorem 3.21, less than $C_1(\text{hypotenuse})^2 < 4C_1 r^2$, hence $f(r)\alpha < 4C_1 r^2$; therefore $f(r) < (24/\pi)C_1 r^2$.

Lemma 3.5: There is a positive constant L such that if $r < L$, then

$$\frac{g(r)}{r} < \frac{6}{\pi}, \qquad \frac{f(r)}{r^2} < \frac{24}{\pi}C_1,$$

where C_1 is the constant in Theorem 3.21. Also, $f(r)$ and $g(r)$ are increasing functions of r.

Chapter 4

$\mathbb{H}^3$ and Euclidean Approximations in $\mathbb{H}^2$

To investigate the fine structure of hyperbolic geometry, we use a method due essentially to Lobachevski. It turns out that the geometry of the horosphere, a certain surface in three-dimensional hyperbolic space $\mathbb{H}^3$, is Euclidean if the definitions are made as follows. The "straight lines" of that geometry are the horocycles in that surface, the "distances" are the arclengths between points along horocycles lying in that surface, and the "angle" between two intersecting horocycles is the angle between the tangent vectors. In that sense, the horosphere, together with the geometry that it inherits from the $\mathbb{H}^3$ in which it is embedded, is a *model* of the Euclidean plane. In it, parallel horocycles are equidistant throughout their entire lengths; there are Cartesian coordinates; and so on. All the formulas and relations of Euclidean geometry apply. Therefore, by projecting small figures in that surface onto a tangent plane in $\mathbb{H}^3$, we find that in that plane, for small figures near the point of tangency, the Euclidean formulas hold approximately, with relative error terms that tend to zero as the sizes of the figures tend to zero. In order to carry out this approach, we first develop three-dimensional hyperbolic geometry, which, however, has independent interest. We give a set of axioms, which, like those of the hyperbolic plane, are user friendly. The first seven axioms are almost identical with those of the hyperbolic plane and the last three deal with planes in $\mathbb{H}^3$. We then develop a set of theorems of the geometry of $\mathbb{H}^3$, and we prove that the geometry inherited by a horosphere is Euclidean.

4.1 Summary

As stated earlier, the clue to quantitative relations (actual formulas) is the locally almost Euclidean nature of hyperbolic geometry. Many Euclidean formulas hold approximately for small figures, with relative error terms that tend to zero as the sizes of the figures tend to zero. For example, the Euclidean formulas $a = c \sin \alpha$ and $b = c \cos \alpha$ for the right triangle in Fig. 2.5f are replaced by

$$a = (c \sin \alpha)(1 + \varepsilon), \qquad b = (c \cos \alpha)(1 + \varepsilon'). \qquad (4.1\text{-}1)$$

where ε and ε' are quantities that tend to zero as $c \to 0$. Similar formulas hold for the law of sines, the law of cosines, the Pythagorean formula, and the law of similarity. For the law of similarity, if a', b', c' are the corresponding side lengths of another right triangle with the same value α of one of the acute angles, then, first, the other acute angles are almost equal and so are the ratios of side lengths:

$$\frac{a'}{b'} = \frac{a}{b}(1 + \varepsilon_1), \quad \frac{b'}{c'} = \frac{b}{c}(1 + \varepsilon_2), \quad \frac{c'}{a'} = \frac{c}{a}(1 + \varepsilon_3), \qquad (4.1\text{-}2)$$

where the ε_i are quantities that tend to zero as $\max(c, c')$ tends to zero.

By means of such formulas and taking certain limits, it is possible to derive differential equations for various functions and thus obtain the formula for the angle of parallelism, detailed formulas for (not necessarily small) triangles, and so on. Some authors derive the final formulas from one of the models of the hyperbolic plane, but we take the attitude that one is not justified in using a model until the categoricalness of the axioms has been established, so that a formula obtained from one model holds for other models, too. (Needless to say, the proof of categoricalness must not depend on anything obtained from a model.)

The derivation of (4.1-1) and (4.1-2) and similar formulas from the axioms of the hyperbolic plane can be accomplished in various ways. We shall take advantage of an idea due essentially to Lobachevski involving three-dimensional hyperbolic geometry (which is derived below). We shall also use the methods of differential geometry.

4.2 Axioms for $\mathbb{H}^3$

The hyperbolic 3-space is a set of otherwise undefined elements called points, denoted by $A, B, C, \ldots$, together with certain subsets called planes, denoted by $\mathcal{P}, \mathcal{Q}, \mathcal{R}, \ldots$, and certain other subsets called lines, denoted by $\ell, m, n, \ldots$, subject to the axioms given below. (It will follow from the axioms that each line is a subset of at least one plane.)

Axioms 1, 2, 3, and 6 are exactly the same as for the plane case, and axioms 4, 5, and 7b are nearly the same.

Axiom 1: Any two points determine (lie in) a unique line.

Axiom 2: A distance function $|AB|$ is defined for each pair of points; it satisfies the triangle inequalities as in the plane case (see Chapter 1).

Axiom 3: A one-to-one correspondence can be established between the set of points in a line and the set $\mathbb{R}$ of real numbers in such a way that if A

and B are any points on the line, then $|AB| = |x(A) - x(B)|$, where $x(A)$ and $x(B)$ are the numbers that correspond to the points A and B.

Axiom 4: If a line lies in a plane, then it separates that plane into two half-planes, in accordance with Axiom 4 of the plane case.

Rays, angles, and triangles are defined as in the plane case.

Axiom 5: To each angle is associated a number, its *radian measure*, in such a way that the angles that lie in a given plane satisfy Axiom 5 of the plane case.

Axiom 6: The side-angle-side criterion for congruence of triangles holds, as in the plane case. *Note*: It will appear that any triangle lies in a plane, but here it is not assumed that the two triangles being compared lie in the *same* plane.

Axiom 7b: There exists a plane, a line in the plane, and a point in the plane not on the line, such that through the point at least two lines can be drawn in the plane that do not intersect the given line.

Axiom 8: For any two distinct points in a plane, the line that joins them (by Axiom 1) lies entirely in the plane.

Axiom 9: Any three noncollinear points lie in a unique plane.

Axiom 10: Given any plane $\mathcal{P}$, there are two subsets HS_1 and HS_2 of $\mathbb{H}^3$, called the *half-spaces bordered by* $\mathcal{P}$, such that the subsets HS_1, HS_2, $\mathcal{P}$ are disjoint and their union is all of $\mathbb{H}^3$, and (1) if points A, B are in the same half-space, the segment AB does not intersect $\mathcal{P}$, (2) if points A, B are in opposite half-spaces, the segment AB intersects the plane $\mathcal{P}$ (in a single point).

Remark: The axioms of $\mathbb{H}^2$ hold in any plane in $\mathbb{H}^3$.

4.3 Some Neutral Theorems of 3-Space

The first two theorems follow directly from Axioms 1, 8, and 9.

Theorem 4.1: A line and a point not on it determine (lie in) a unique plane.

Theorem 4.2: Two distinct intersecting lines determine (lie in) a unique plane.

Theorem 4.3: If two distinct planes intersect, the intersection is a line.

Proof: First, if the intersection contains at least two points, then, by Axiom 8, the line through those points lies in both planes, hence in the intersection, and the intersection is just that line, for if it contained any other point, then, by Theorem 4.1, the planes would coincide, hence would not be distinct. Therefore, it suffices to show that the intersection cannot be just a single point (as can happen in 4-space). Let Z be a point of the intersection and let AB be a line segment in the first plane that contains Z in its interior. If AB contains any other point of the intersection, we are finished. Otherwise, points A and B are on opposite sides of the second plane. Let C be any other point in the first plane not on the line containing A and B. If the segment AC contains a point of intersection, we are again finished; otherwise A and C are on the same side of the second plane; hence C and B are on opposite sides of the second plane; hence, by Axiom 10, the segment BC intersects the second plane, but it lies in the first plane and this gives a second point of intersection, as required.   □

In the following, we denote by $\mathcal{P}(A, B, C)$ the unique plane determined by noncollinear points A, B, C, by $\mathcal{P}(\ell, A)$ the unique plane determined by a line ℓ and point A not on ℓ, and by $\mathcal{P}(\ell, m)$ the unique plane determined by distinct intersecting lines ℓ, m.

Definition 4.1: If a line m intersects a plane $\mathcal{P}$ at point A, we say that m is *perpendicular to* $\mathcal{P}$ if m is perpendicular to every line in $\mathcal{P}$ that passes through the point A.

Comment. It is easy to see that $\mathcal{P}$ contains *all* lines ℓ that are perpendicular to m at A, for if ℓ is any such line, the perpendicular to m at A in the plane $\mathcal{P}(\ell, m)$ is unique.

Theorem 4.4 ("Dropping a perpendicular"): If A is a point not in a plane $\mathcal{P}$, there is a unique line m through A perpendicular to $\mathcal{P}$.

Proof: Let $f(X)$ denote the distance $|AX|$ from the point A to a variable point X in $\mathcal{P}$. As in the proof of Theorem 2.2 one can prove that $f(X)$ is a continuous function of the point X in the plane $\mathcal{P}$, and $f(X)$ approaches infinity as X moves off to infinity on any ray. Therefore, by basic properties of continuous functions defined in a plane, there is a point B in the plane where $f(X)$ attains a minimum value. (See the Exercise below.) Then, along any line ℓ in $\mathcal{P}$ through B the function has a minimum value at B; hence, by Theorem 2.2, the line AB is perpendicular to the line ℓ in the plane $\mathcal{P}(\ell, A)$. Uniqueness follows as in the theorem just referred to: the assumption of two perpendiculars to $\mathcal{P}$ through A would lead to a contradiction.   □

If B is the intersection of the line m and the plane $\mathcal{P}$ of the theorem, then the *distance of A from $\mathcal{P}$* is defined to be $|AB|$.

Corollary 1: If a plane and a sphere intersect as in Fig. 4.3a, the intersection is either a point (i.e., in case of tangency) or a circle.

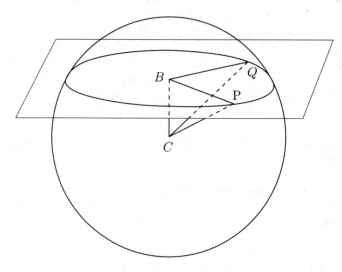

Fig. 4.3a.

Proof: Let C be the center of the sphere and r its radius. If C lies in the plane $\mathcal{P}$, the result is obvious, so suppose it doesn't. Drop a perpendicular from C to $\mathcal{P}$, with foot at B. If $|CB| = r$, then B is a point of the sphere, and, since a leg of a right triangle is shorter than the hypotenuse, all other points of the plane are at distance from C greater than r; hence the intersection consists of the single point B. If $|CB| < r$, let P, Q be any two points of the intersection. Then, by the right angle-side-side criterion for congruence of right triangles, the triangles BCP and BCQ are congruent; hence the lengths $|BP|$ and $|BQ|$ are equal; hence the intersection is a circle, as required. $\qquad\square$

Corollary 2: If lines ℓ_1 and ℓ_2 are both orthogonal to a line m at point A, *then m is perpendicular to the plane* $\mathcal{P} = \mathcal{P}(\ell_1, \ell_2)$.

Proof: (See Fig. 4.3b.) What has to be shown is that if ℓ is any other line through A in $\mathcal{P}$, then ℓ is also perpendicular to m. Let C be a point on m not in the plane, and let r be a length greater than $|AC|$. Consider the intersection of a sphere of radius r and center C with the plane; according to the preceding corollary, that intersection is a circle. Let Q_1, Q_1' be the

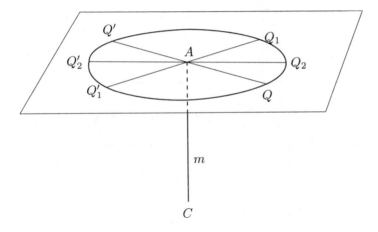

Fig. 4.3b.

intersections of ℓ_1 with the circle and Q_2, Q'_2 those of ℓ_2. By hypothesis, the angles $\angle CAQ_1, \angle CAQ'_1, \angle CAQ_2$, and and $\angle CAQ'_2$ are all right angles. By the right angle-side-side criterion, again, the triangles CAQ_1, CAQ'_1, CAQ_2, and CAQ'_2 are all congruent, hence $|AQ_1|, |AQ'_1|, |AQ_2|$, and $|AQ'_2|$ are all equal, and it follows that A is the center of the circle. Now let ℓ be any other line in the plane through the point A, and let Q, Q' be its intersections with the circle. Since $|AQ| = |AQ_1|$ and $|CQ| = |CQ_1|$, the side-side-side criterion shows that the triangles CAQ and CAQ_1 are congruent. Therefore, the angle $\angle CAQ$, like $\angle CAQ_1$ is a right angle; hence ℓ is perpendicular to m, as was to be proved.   □

Theorem 4.5 ("Erecting a perpendicular"): If A is a point in a plane $\mathcal{P}_0$, there is a line m perpendicular to $\mathcal{P}_0$ at A.

Proof: Let ℓ_1 and ℓ_2 be mutually perpendicular lines through A and lying in $\mathcal{P}_0$, as in Fig. 4.3c. Let B be a point not in $\mathcal{P}_0$, and let $\mathcal{P}_1$ be the plane $\mathcal{P}(\ell_1, B)$. Let ℓ_3 be the line in that plane perpendicular to ℓ_1 at A. Let $\mathcal{P}_2$ be the plane $\mathcal{P}(\ell_2, \ell_3)$, and let m be the line in this last plane perpendicular to ℓ_2 at A. We claim that m is the required line. Since ℓ_1 is perpendicular to ℓ_2 and ℓ_3, it is perpendicular to the plane $\mathcal{P}_2$, hence ℓ_1 is also perpendicular to m. Then, since m is perpendicular to both ℓ_1 and ℓ_2, it is perpendicular to $\mathcal{P}_0$, as was to be proved.   □

Definition 4.2: Let ℓ be the line of intersection of planes $\mathcal{P}_1$ and $\mathcal{P}_2$, and let Z be a point of the line ℓ. In each of the planes, choose one of the half-planes bordered by ℓ. The *angle α between the half-planes at Z* is defined

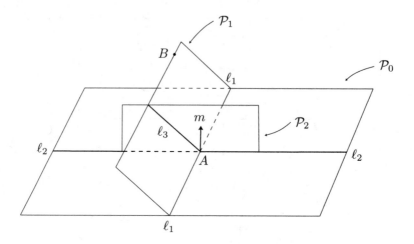

Fig. 4.3c.

as follows: In each of the half-planes, construct a ray from Z perpendicular to ℓ; then α is defined as the angle between those two rays.

Theorem 4.6: For the given half-planes, the angle α in Definition 4.2 is independent of the choice of the point Z on the line ℓ.

   Note: For this reason, the angle α is called simply the *angle between the half-planes*. Since each of the planes $\mathcal{P}_1$ and $\mathcal{P}_2$ has two half-planes bordered by ℓ, the two planes determine four such angles, which are equal in pairs and supplementary in pairs.

Proof: Let A and C be any two points on the line ℓ of intersection of planes $\mathcal{P}_1$ and $\mathcal{P}_2$. We must show that the angles labeled α and β in Fig. 4.3d are equal. Let B be the midpoint of AC. In $\mathcal{P}_1$ draw a line DG through B and in $\mathcal{P}_2$ a line EF through B so that both these lines make the same angle with the line ℓ. Since the angles $\angle BAD, \angle BAE, \angle BCF$, and $\angle BCG$ are all right angles and, in each plane, the vertical angles at B are equal, the angle-side-angle criterion shows that the triangles ABD, ABE, BCF, and BCG are all congruent. Therefore, $|AD| = |AE| = |CF| = |CG|$ and $|BD| = |BE| = |BG| = |BF|$. From the latter it follows from the SAS criterion that the triangles BDE and BFG are congruent, so that $|DE| = |FG|$, and then it follows from the former by the SSS criterion that the triangles ADE and CFG are congruent. Therefore, the angles $\angle DAE$ and $\angle FCG$ are equal, so that $\alpha = \beta$, as required, since $\angle FCG$ and β are vertical angles and hence equal. $\square$

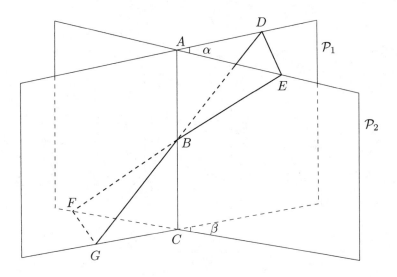

Fig. 4.3d.

Corollary: Two lines perpendicular to a plane are coplanar.

Proof: Let ℓ_1 and ℓ_2 be perpendicular to $\mathcal{P}_0$ at B_1 and B_2, respectively. See Fig. 4.3e. Let m be the line containing B_1 and B_2 (it lies in $\mathcal{P}_0$) and let n be the line in $\mathcal{P}_0$ perpendicular to m at B_2. (*Note:* The line ℓ_2 is not shown in the drawing, but we shall soon find it.) In the plane $\mathcal{P}_1$ of ℓ_1 and B_2 (which is perpendicular to $\mathcal{P}_0$ because ℓ_1 is), erect the perpendicular k to m at B_2. Then k is also perpendicular to n, because of Theorem 4.6. It follows that k is perpendicular to $\mathcal{P}_0$, hence it coincides with ℓ_2. Therefore, ℓ_1 and ℓ_2 both lie in the plane $\mathcal{P}_1$, as required. See Fig. 4.3e.   □

Exercise
1.   Give details of the proof that the function f used in the proof of Theorem 4.4 assumes a minimum value. *Hint:* Choose a point Z and a ray starting from Z, and introduce polar coordinates in the plane $\mathcal{P}$. Prove that there is a positive real number R with the property that if the distance from a point X in $\mathcal{P}$ to Z is greater than R, then $f(X) > 2f(Z)$. Then f corresponds to a continuous function on the closed rectangle $[0, R] \times [0, 2\pi]$, whose minimum value is the minimum value of f.

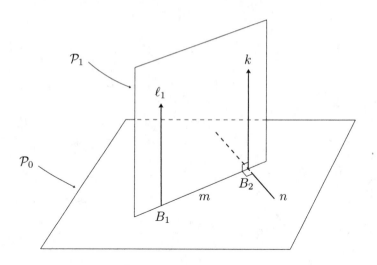

Fig. 4.3e.

4.4 Spherical Coordinates

We define spherical coordinates r, ϕ, θ in $\mathbb{H}^3$ in analogy with the polar co-ordinates r, θ that were defined in $\mathbb{H}^2$ in Section 3.4.

We let Z be an arbitrary point, called the *origin*; we let ℓ be a line through the origin called the *polar axis*; we let $\overrightarrow{\ell}$ be one of the rays in ℓ from Z, called the *north polar axis*; we let HP_0 be a half-plane bordered by ℓ called the *half-plane of the prime meridian,* as in Fig. 4.4a; and we let HS be one of the half-spaces bordered by the plane containing HP_0, called the *western half space.* Then, if P is any point in $\mathbb{H}^3$, we define r as the distance $|ZP|$ and ϕ (the colatitude) as the angle between the ray $\overrightarrow{ZP}$ and the north polar axis $\overrightarrow{\ell}$; and, for P in the western half-space HS, we define θ (the longitude) as the angle between the half-plane HP_0 and the half-plane bordered by ℓ and containing P; and for P in the opposite or *eastern* half-space, we define θ to be minus the angle between those two planes. These definitions are understood to be subject to the following restrictions:

(1) If $r = 0$, ϕ and θ are undefined;
(2) If $r > 0$ and ϕ is 0 or π, then θ is undefined;
(3) $r \geq 0$, $0 \leq \phi \leq \pi$, $-\pi < \theta \leq \pi$.

With these understandings, there is a one-to-one correspondence between triples (r, ϕ, θ) and the points of $\mathbb{H}^3$.

Although the definition of the spherical coordinates is the same as for Euclidean 3-space, the coordinates behave differently in the two cases. For

Fig. 4.4a.

example, while the equation of a plane perpendicular to the polar axis is $r\cos\phi = $ const in the Euclidean case, we shall see in Chapter 9 that it is $\tanh r\cos\phi = $ const in the hyperbolic case.

4.5 Isometries in $\mathbb{H}^3$

As in the two-dimensional case, a one-to-one mapping σ of $\mathbb{H}^3$ onto itself is an *isometry* if it preserves all distances; that is, if for any two points P, Q and their images under σ, P', Q', we have $|PQ| = |P'Q'|$.

An isometry σ has the following fundamental properties: (1) It follows from the SSS criterion for congruence of triangles that σ also preserves all angles. (2) As noted in Chapter 2, points A, B, C lie on a line if and only if one of the inequalities (less than or equal) in the triangle inequality (1.1-1) is equality, and it follows, since distances are preserved, that lines are mapped onto lines by σ. (3) Since, according to Section 4.2 above, a plane in $\mathbb{H}^3$ is the set of all points on lines ℓ perpendicular to a line m at a point A of m and since right angles are mapped onto right angles, it follows that σ maps planes onto planes.

Let r, ϕ, θ be spherical coordinates as above, and let r', ϕ', θ' be another set obtained from a different choice of $Z, \ell, \overrightarrow{\ell}, HP_0$, and HS. Let σ be the mapping obtained by mapping a point P with coordinates r, ϕ, θ, onto a point P' with coordinates having the same values in the second system, $r' = r$, $\phi' = \phi$, and $\theta' = \theta$. This is obviously a one-to-one mapping of $\mathbb{H}^3$ onto itself.

Theorem 4.7: The mapping σ just defined preserves distances, hence is an isometry; that is, if P_1, P_2 are any points and P_1', P_2' are their images under σ, then $|P_1 P_2| = |P_1' P_2'|$.

Proof: We consider the case in which both points are in the "northern" half-space, that is, r_1 and r_2 are positive and the angles ϕ_1 and ϕ_2 are less than a right angle. Essentially the same argument works if both points are in the "southern" half-space. It is evident what to do if one point (or both) is on the polar axis or in the equatorial plane. The necessary modification when one point is in the northern half-space and the other is in the southern is left to the reader. From each of the points, P_1 and P_2 we drop perpendiculars to the equatorial plane (the plane $\phi = \pi/2$), and we call B_1 and B_2 the feet of those perpendiculars. Since the angle $\angle P_1 Z B_1$ is the complement of ϕ_1, we see from the side-angle-angle criterion that the triangles $P_1 Z B_1$ and $P_1' Z' B_1'$ are congruent; hence the lengths of the segments $Z B_1$ and $P_1 B_1$ are preserved under the mapping. The same is true of the segments $Z B_2$ and $P_2 B_2$. Then, since the angle between $Z B_1$ and $Z B_2$ is $|\theta_2 - \theta_1|$, the triangle $B_1 Z B_2$ and its image under σ (these are triangles in the two equatorial planes) are congruent, and it follows that the distances $|B_1 B_2|$ and $|B_1' B_2'|$ are equal. According to the corollary to Theorem 4.6, the perpendiculars lie in a plane; hence we consider the quadrilateral $P_1 B_1 B_2 P_2$ and its image under the mapping. The angles at B_1 and B_2 are right angles and the lengths of $P_1 B_1, B_1 B_2$, and $B_2 P_2$ are preserved, and we see from the SASAS criterion (see Exercise 1 in Section 2.8) that that quadrilateral and its image are congruent; hence the distance $|P_1 P_2|$ is preserved under the mapping, as was to be proved. $\qquad\square$

We note that, as in the plane case, *any* isometry can be presented presented in this way: It is only necessary to choose $Z, \ell, \overrightarrow{\ell}, HP_0, HS$ arbitrarily and then to take $Z', \ell', \overrightarrow{\ell}', HP_0', HS'$ to be their images under the isometry.

Special cases of isometry are the following: a translation T_d by a distance d along the polar axis, a rotation R_α^z by angle α about the polar axis (the z-axis), a rotation R_β^x by angle β about the x-axis (the line where $\phi = \pi/2, \theta = 0$), and a reflection M obtained by changing the sign of θ. Given a general isometry as described above in terms of two spherical coordinate systems, we can undo the mapping (find its inverse) by making first a rotation R_α^z (if necessary) to make the second origin Z' lie in the y-z plane of the first system (where $\theta = \pm\pi/2$), then making a rotation R_β^x to bring Z' onto the polar axis, then making a translation T_d to bring Z' to the origin Z, then making a rotation R_γ^z to bring $\overrightarrow{\ell}$ into the y-z plane, then a rotation R_δ^x to bring $\overrightarrow{\ell}'$ into coincidence with $\overrightarrow{\ell}$, then a rotation R_ε^z to bring the half-plane HP_0' into coincidence with HP_0, and finally, making a reflection M in the x-z plane, if necessary, to bring the half-space HS'

into coincidence with HS. Since the inverse of an arbitrary isometry is an arbitrary isometry, we have proved the following:

Theorem 4.8: The general isometry can be written in one of the forms

$$R_\varepsilon^z R_\delta^x R_\gamma^z T_d R_\beta^x R_\alpha^z \quad \text{or} \quad M R_\varepsilon^z R_\delta^x R_\gamma^z T_d R_\beta^x R_\alpha^z \qquad (4.5\text{-}1)$$

The first is a direct isometry, the second a reverse isometry.

We see that the isometry group has six parameters.

Remark: Although the factors in the above expressions don't in general commute, they can be written in different orders, at the expense of changing the values of the parameters $\varepsilon, \delta, \gamma, d, \beta, \alpha$. For example, if σ is a given isometry and its inverse σ^{-1} is written as above, then σ will appear with the factors in the opposite order with the signs of the subscripts reversed. If σ is a reverse isometry and we write $\sigma = M\mu M$, where μ is as above, then σ has M at the right.

4.6 Asymptotic Bundles and Ideal Points at Infinity

Let ℓ and m be directed lines in $\mathbb{H}^3$; ℓ is *asymptotic* to m if the distance of a point P on ℓ to m decreases to zero as the the coordinate $x(P)$ goes to infinity. It is easy to see that then m is also asymptotic to ℓ. (We assume that the lines are directed in the same sense, by which is meant that as $x(P)$ increases, the coordinate of the foot of the perpendicular from P to m also increases.) Also, if two lines are asymptotic to a third, they are asymptotic to each other. The set of all lines asymptotic to a given line constitute an *asymptotic bundle*. The lines of a bundle fill up all of $\mathbb{H}^3$, because if any point and any line are given, they determine a plane. Hence, according to Chapter 3 there is always a line in that plane through the given point asymptotic to the given line.

Lemma 1: Any two asymptotic lines are coplanar.

Proof: If ℓ and m are asymptotic, let A be a point on ℓ, and let $\mathcal{P}$ be the plane determined by A and m. Let ℓ' be the line in $\mathcal{P}$ through A asymptotic to m, and suppose $\ell' \neq \ell$. Then ℓ and ℓ' lie in a plane $\mathcal{P}'$ and intersect at A. According to Exercise 4 in Section 2.8, the distance from ℓ to ℓ' increases as the points move to infinity along them, but that would contradict the fact that they are asymptotic (because they are both asymptotic to m). Therefore ℓ and ℓ' coincide, so that ℓ and m are coplanar, as was to be proved. □

Lemma 2: Let r, ϕ, θ be spherical coordinates. Asymptotic lines have the same limiting values of ϕ, θ as the points on them move to infinity in the direction of the lines, unless the limiting value of ϕ is 0 or π, in which case the values of θ need not converge.

The proof is omitted; it is similar to the proof of the corresponding theorem in the plane case.

Definition 4.3: The same is true of all the lines of an asymptotic bundle. With each bundle, we associate an *ideal point* at infinity, having the coordinates (ϕ_0, θ_0) where ϕ_0 and θ_0 are the limiting values of ϕ and θ, it being understood that if ϕ_0 is 0 or π, then θ_0 is irrelevant.

Definition 4.4: Let $\mathcal{P}$ and $\mathcal{Q}$ be distinct nonintersecting planes. If there exist asymptotic directed lines ℓ and m, one lying in each of the planes, we say that the planes are *asymptotic* in the direction of the ideal point determined by those lines, or in the corresponding *ideal direction*.

Theorem 4.9: Two distinct and nonintersecting planes cannot be asymptotic in more than one ideal direction.

Proof (Indirect): Suppose two planes were asymptotic in each of two ideal directions. In each of the planes, those directions would determine ideal points in the sense of the preceding chapters, and in each plane there would be a line joining those ideal points. Those lines would be asymptotic (in fact, in both directions), so the distance between them would tend to zero in both directions, and they would in turn lie in a plane (because they are asymptotic) but the distance between two lines in a plane cannot tend to zero in both directions, unless the lines coincide, but then the given planes would not be nonintersecting.   □

Theorem 4.10: Given a plane $\mathcal{P}_0$, a directed line ℓ_0 in $\mathcal{P}_0$ and a point A not in the plane, there is one and only one plane $\mathcal{P}_1$ that contains A and is asymptotic to $\mathcal{P}_0$ in the direction of the line ℓ_0.

Proof: Drop a perpendicular from A to $\mathcal{P}_0$ and denote by B the foot of the perpendicular. Let ℓ_1 be the unique line in $\mathcal{P}_0$ through B and asymptotic to ℓ_0. Let $\mathcal{P}_2$ be the plane containing ℓ_1 and A, and let $\mathcal{P}_3$ be the plane perpendicular to ℓ_1 at B, as in Fig. 4.6a. Let ℓ_2 be the unique line in $\mathcal{P}_2$ through A asymptotic to ℓ_1 and let ℓ_3 be the line in $\mathcal{P}_3$ perpendicular to AB. We claim that the plane through the lines ℓ_2 and ℓ_3 is the desired plane $\mathcal{P}_1$. (Uniqueness is discussed below.) Since $\mathcal{P}_0$ and $\mathcal{P}_1$ contain the asymptotic lines ℓ_1 and ℓ_2, respectively, it only remains to show that those planes don't intersect. Clearly, they don't intersect at any point of ℓ_1. Since the entire figure is invariant under reflection in the plane $\mathcal{P}_2$, we see that if

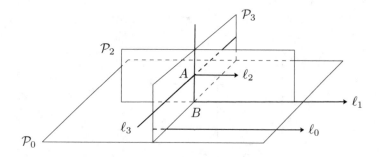

Fig. 4.6a.

there were a point of intersection on one side of the line ℓ_1, there would be
another point of intersection on the other side, and the line joining those
two points of intersection in $\mathcal{P}_0$ would give a point of intersection on ℓ_1,
which we have just shown cannot happen. Hence $\mathcal{P}_1$ is asymptotic to $\mathcal{P}_0$ in
the direction of ℓ_1, which is the same as the direction of the original line
ℓ_0, because ℓ_1 and ℓ_0 are asymptotic. We now show uniqueness. For that
purpose, it is slightly more convenient to interchange the roles of $\mathcal{P}_0$ and
$\mathcal{P}_1$ and to show that $\mathcal{P}_0$ is the only plane through point B asymptotic to
$\mathcal{P}_1$ in the direction of the line ℓ_2. Suppose that there were another such
plane $\mathcal{P}_0'$ containing a line ℓ_1' asymptotic to ℓ_2. Then, ℓ_1 and ℓ_1' would be
asymptotic to each other, but they both pass through the point B; hence
they would coincide. Hence they are just the line of intersection of the two
planes $\mathcal{P}_0$ and $\mathcal{P}_0'$. We now move plane $\mathcal{P}_3$ to the right, in the direction of ℓ_1
and ℓ_2, keeping it perpendicular to the line ℓ_1. In that process, the angle of
intersection of $\mathcal{P}_0$ and P_0' is a constant α, according to Theorem 4.6, while
the distance $|AB|$ tends to zero. The situation is then as in Fig. 4.6b where
ℓ_4 and ℓ_4' are the lines of intersection of the planes $\mathcal{P}_0$ and P_0' with $\mathcal{P}_3$.
As soon as the distance $|AB|$ has become small enough that the angle
of parallelism $\Pi(|AB|)$ differs from 90° by less than one-half of α, then at
least one of the lines ℓ_4 and ℓ_4' would intersect the plane $\mathcal{P}_1$, and that would
contradict the requirement that neither of the planes $\mathcal{P}_0$ or $\mathcal{P}_0'$ intersects
$\mathcal{P}_1$. That completes the proof.   □

Exercises

1. Show that if two planes are nonintersecting and nonasymptotic, then
    they have a unique common perpendicular. Such planes are called *ul-
    traparallel*.
2. Show that if planes $\mathcal{P}_1$ and $\mathcal{P}_2$ are distinct and asymptotic, then there
    is a unique plane perpendicular to both planes. *Hint:* If ℓ_1 and ℓ_2 are
    asymptotic lines in $\mathcal{P}_1$ and $\mathcal{P}_2$ respectively, then we can parametrize
    the lines in $\mathcal{P}_1$ that are asymptotic to ℓ_1 in the same direction as ℓ_2

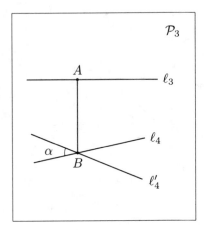

Fig. 4.6b.

is by a horocycle C_1 that is orthogonal to all those lines. We can do the same for the lines asymptotic to ℓ_2 in P_2 with a horocycle C_2. These horocycles are parametrized by $\mathbb{R}$, as in Section 3.9. Let x be a coordinate along C_1, and y a coordinate along C_2. Then each ordered pair (x, y) of real numbers has an associated pair of asymptotic lines and hence the plane $P_{x,y}$ determined by those two lines. Investigate the angles between $P_{x,y}$ and the planes P_1 and P_2 as functions of (x, y).

4.7 The Horosphere; The Coordinates ξ, η, ζ; Ideal Rotations

The horosphere may be described in either of two ways. First, the circle in Fig. 3.9a is replaced by a sphere with center at Q and tangent at P to the plane perpendicular to the line ℓ at P. As the point Q moves off to infinity along the line ℓ keeping P fixed, the sphere converges to a surface called a *horosphere*. Second, we consider a horocycle perpendicular to a line ℓ and rotate it around the line ℓ, thus generating the horosphere.

In Fig. 4.7a (which is the analogue of Fig. 3.9c), we have an asymptotic bundle of lines and a family of horospheres orthogonal to them. Any two lines of the bundle determine a plane, and the lines of the bundle that lie in that plane form an asymptotic pencil in that plane and are all perpendicular to each of the horospheres. In that sense, any such plane is orthogonal to each of the horospheres. Hence, an alternative interpretation of Fig. 4.7a is that the lines in the figure (at $\eta = $ const) represent an asymptotic sheaf of

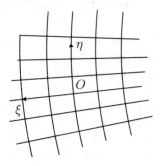

Fig. 4.7a.

planes normal to the plane of the drawing and intersecting the horospheres normally. Any two planes of the sheaf are asymptotic, as defined in Section 4.6. The entire structure is invariant under reflection in any of those planes, and the resultant of two such reflections is an ideal rotation under which each plane of the sheaf is mapped onto another plane of the sheaf and the horospheres are mapped onto themselves. Then there is another asymptotic sheaf of planes normal to those of the first sheaf (the plane of the drawing is one) and asymptotic in the direction of the same ideal point at infinity as the first sheaf (to the right in the drawing).

We now define coordinates ξ, η, ζ in analogy with the coordinates ξ, η of Section 3.9. The value of η is constant on each of the planes of the first sheaf mentioned, and ζ is constant on each of the planes of the second sheaf. An arbitrary origin O is chosen; η and ζ are measured in the horosphere of the family that contains the point O (called the *principal* horosphere). The value of η is the arclength from the point O along the horocycle in which the plane of the drawing intersects the principal horosphere, and then ζ is the arclength in the plane at coordinate η along along the horocycle perpendicular to the plane of the drawing. Lastly, ξ is distance along the line of the bundle having coordinates η, ζ. (We leave it to the reader to decide about the signs of those coordinates.)

We now consider isometries in $\mathbb{H}^3$ obtained by adding constants to ξ, η, ζ. Adding a constant to ξ, keeping η, ζ fixed, gives a translation along the line of the bundle that passes through the origin. Adding a constant to η, keeping ξ, ζ fixed, is an ideal rotation in which each plane of the first sheaf is mapped onto another, and adding a constant to ζ, keeping ξ, η fixed, is a similar ideal rotation associated with the second sheaf. These, taken together, constitute a 3-parameter subgroup of the isometry group under which the ideal point associated with the asymptotic bundle is fixed. Rotations about the line of the bundle that passes through the origin also keep that ideal point fixed.

4.8 The Euclidean Geometry in a Horosphere

In this section, we set $\xi = 0$, in the notation of the preceding section; we are thus dealing with the horosphere passing through the origin O. We shall see that η and ζ behave in all respects like Cartesian coordinates in that horosphere. By *distance* in the horosphere between two of its points, we mean the arclength along a horocycle that joins the points. By *lines* in the horosphere we mean the horocycles lying in it, and by the *angle* between two of those lines at a point where they intersect we mean the angle between their tangent vectors, which is equal to the angle between the corresponding planes orthogonal to the horosphere.

Since adding a constant to ζ is an isometry (an ideal rotation), we see that the distance (arclength) between points η_1, ζ and η_2, ζ is the same as the distance between $\eta_1, 0$ and $\eta_2, 0$. That is, the lines (horocycles) $\eta = $ const are parallel in the Euclidean sense. The same applies to the lines $\zeta = $ const. By use of these ideal rotations, we see also that the angle of intersection of any lines $\eta = $ const and $\zeta = $ const is the same as the angle of intersection at the origin of the lines $\eta = 0$ and $\zeta = 0$, namely $90°$. Therefore, the coordinates η, ζ are Cartesian, and the geometry is Euclidean. We say that this is the geometry that the horosphere *inherits* from the $\mathbb{H}^3$ in which it lies. (For further treatment of the notion of an inherited geometry, see the next chapter.)

In particular, the Pythagorean law holds in the horosphere: if a and b are the legs of a right triangle and c is the hypotenuse, then $a^2 + b^2 = c^2$.

4.9 Euclidean Fine Structure of the Hyperbolic Plane

For projection from a horosphere onto a tangent plane, we shall need the two lemmas proved below.

By power series expansions of the functions $\tan \alpha$ and $\sec \alpha$, we find for the right triangle in Fig. 4.9a in the Euclidean case that for fixed b and small α, a is $b\alpha$ plus higher terms and c is $b(1 + \alpha^2/2)$ plus higher terms. A similar thing holds in the hyperbolic case, within certain bounds.

Lemma 1: For the right triangle of Fig. 4.9a in $\mathbb{H}^2$ there are constants C_1, C_2, b_0, α_0 such that if $b < b_0$ and $\alpha < \alpha_0$, then

$$a < C_1 b \alpha < C_1 c \alpha \quad \text{and} \quad c - b < C_2 b \alpha^2 < C_2 c \alpha^2. \qquad (4.9\text{-}1)$$

(In each case, the second inequality follows from the first because $b < c$.)

Proof: Choose b_0 arbitrarily and let α_0 be the acute angle of an isosceles right triangle with legs of length b_0 ($\alpha_0 < 45°$). Bisect the angle α_0 at

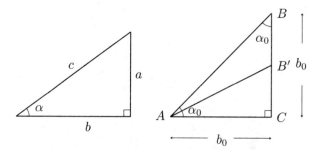

Fig. 4.9a   Fig. 4.9b

A as in Fig. 4.9b and set $\alpha_1 = \frac{1}{2}\alpha_0$, $b_1 = |B'C|$. It is easy to see that $|B'C| < |BB'|$ (drop a perpendicular from B' to AB and use the SAA criterion for congruence of the resulting triangles with AB' as hypotenuse), so that $b_1 < \frac{1}{2} b_0$. We then bisect the angle α_1 at vertex A of the triangle $AB'C$, we define α_2 and b_2 in similar fashion, and so on. In this way we obtain a sequence of thinner and thinner triangles all having AC as one leg. We have

$$\alpha_n = 2^{-n}\alpha_0, \qquad b_n < 2^{-n}b_0.$$

Now consider a right triangle with an angle $\alpha < \alpha_0$ and $b = b_0$. (The case $b < b_0$ is considered below.) Choose n so that $\alpha_{n+1} < \alpha \le \alpha_n$. Denote by a the length of the side opposite vertex A. Then,

$$a < b_n < 2^{-n}b_0 = 2 \cdot 2^{-n-1}b_0 = 2\frac{\alpha_{n+1}}{\alpha_0} b_0 < 2\frac{\alpha}{\alpha_0} b_0 = C_1\alpha b_0,$$

where $C_1 = 2/\alpha_0$. Now suppose that $b < b_0$. We repeat the entire procedure, putting primes on the quantities and setting $b'_0 = b$. Since $b'_0 < b_0$, the new starting triangle fits inside the old one, hence has a smaller angular defect, hence $\alpha'_0 > \alpha_0$, hence $C'_1 < C_1$. Since $\alpha < \alpha_0$, by hypothesis, we have $\alpha < \alpha'_0$, and the new procedure gives that $a' < C'_1\alpha b'_0$, hence $a' < C_1\alpha b'_0$ or, in the original notation, $a < C_1\alpha b$, as required. To establish the other inequality in Equation (4.9-1), we drop a perpendicular from vertex C to the hypotenuse, as in Fig. 4.9c. For $b = |AC|$, we have $|AD| < b < |AB| = c$. For small α, the angular defect of the triangle ADC is small (see Theorem 3.10 in Section 3.2) and it follows that $\angle DCB = \alpha(1 + \varepsilon)$, hence

$$|DB| < C_1|DC|\alpha(1 + \varepsilon) < C_1[C_1 b\alpha]\alpha(1 + \varepsilon),$$

but $c - b$ is less than $|DB|$, and if we replace ε by its greatest value ε_0 for $0 < \alpha < \alpha_0$ and call $C_2 = C_1^2(1 + \varepsilon_0)$, we have

$$c - b < C_2 b\alpha^2,$$

which establishes the other inequality of (4.9-1). This completes the proof of Lemma 1. $\square$

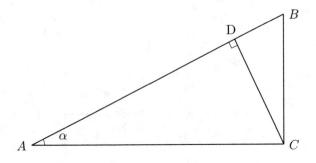

Fig. 4.9c.

We now consider projecting figures in a horosphere $\mathcal{S}$ onto a tangent plane $\mathcal{P}$ by lines of the asymptotic bundle orthogonal to the horosphere, as in Fig. 4.9d. The point P of $\mathcal{S}$ is mapped onto the point Q of $\mathcal{P}$ by ℓ', Z is

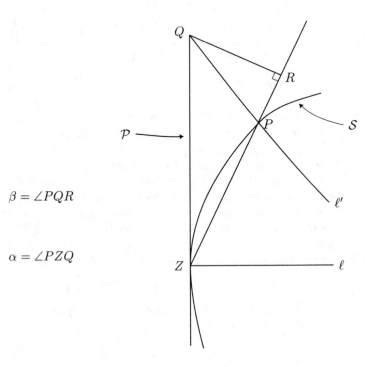

$\beta = \angle PQR$

$\alpha = \angle PZQ$

Fig. 4.9d.

the point of tangency, and the entire structure shown in the figure lies in the plane of the asymptotic lines ℓ and ℓ'. Let s $(= s(P))$ denote the arclength from Z to P along the horocycle shown (which is the intersection of the horosphere S with the plane of the drawing). We wish to show that s and $|ZQ|$ are nearly equal, for points P near Z.

Lemma 2: In the above notation,

$$|ZQ| = s(1 + \varepsilon), \qquad (4.9\text{-}2)$$

where ε is a quantity that $\to 0$ as $s \to 0$.

Proof. It follows from the discussion of arclength in Section 3.9 that the arclength s lies numerically between the length of the chord and the length of the tangent segment ZQ; that is,

$$|ZP| < s < |ZQ|, \qquad (4.9\text{-}3)$$

and we shall show that the ratio $|ZP|/|ZQ|$ approaches 1 as s goes to zero. We drop a perpendicular QR from Q to the line of the chord ZP, as shown. (QP is perpendicular to the horocycle and QR is perpendicular to the line of the chord.) Since the horocycle lies between the tangent and circles with center on the line ℓ and passing through Z, the angle $\alpha = \angle PZQ \to 0$ as $P \to Z$. Then angle $\beta = \angle PQR$ also $\to 0$. We now use the results of Lemma 1, and we shall repeatedly use expressions of the form $1 + \varepsilon$, where ε may represent different quantities on different occasions, but always quantities that $\to 0$ as $s \to 0$. First, since $\alpha \to 0$, $|ZR| = |ZQ|(1+\varepsilon)$. Next, $|QR| = |ZR|\alpha(1 + \varepsilon)$ and $|RP| = |QR|\beta(1 + \varepsilon) = |ZR|\alpha\beta(1 + \varepsilon)$. Hence, since α and $\beta \to 0$ as $s \to 0$, we have $|ZP| = |ZR| - |RP| = |ZR|(1 + \varepsilon)$. Therefore $|ZP| = |ZQ|(1 + \varepsilon)$, and (4.9-2) now follows from (4.9-3), as required.   □

Let $\mathcal{P}$ be a plane in $\mathbb{H}^3$. Recall that the geometry in $\mathcal{P}$ is identical with that of the hyperbolic plane $\mathbb{H}^2$. Let Z be any point of $\mathcal{P}$ and let S be one of the two horospheres tangent to $\mathcal{P}$ at Z. Let ABC be a right triangle in S with right angle at C and with vertex A placed at the point Z of tangency with $\mathcal{P}$. As usual, we call $a = |BC|$, $b = |AC|$, $c = |AB|$, where now $|\cdot|$ denotes distance in S, that is, arclength along horocycle segments, as in the preceding section. See Fig. 4.9a.

Let $A'B'C'$ be the projection of ABC onto $\mathcal{P}$, as described in Sect. 4.1 (with A and A' both at Z) and call $a' = |B'C'|$, $b' = |A'C'|$, $c' = |A'B'|$, where here $|\cdot|$ denotes distance along line segments in $\mathcal{P}$. According to Lemma 2,

$$b' = b(1 + \varepsilon_1), \qquad c' = c(1 + \varepsilon_2), \qquad (4.9\text{-}4)$$

where ε_1 and ε_2 are quantities that $\to 0$ as $c \to 0$. Let $\mathcal{P}_{AC}$ be the plane perpendicular to the horosphere and containing AC, and let $\mathcal{P}_{CB}$ be the plane also perpendicular to the horosphere and containing CB. $\mathcal{P}_{AC}$ is perpendicular to $\mathcal{P}$ and $\mathcal{P}_{CB}$ is perpendicular to $\mathcal{P}_{AC}$. Although $\mathcal{P}_{CB}$ is not perpendicular to $\mathcal{P}$, its intersection with $\mathcal{P}$, namely, the line $C'B'$ is perpendicular to $\mathcal{P}_{AC}$, because it is in two planes ($\mathcal{P}$ and $\mathcal{P}_{CB}$) both perpendicular to $\mathcal{P}_{AC}$. Therefore, the segment $C'B'$ in the plane is perpendicular to the segment $A'C'$ in the plane. That is, the projection $A'B'C'$ is also a right triangle, with right angle at C'. Clearly, the angle $\angle C'A'B'$ is equal to $\angle CAB = \alpha$, because the planes $\mathcal{P}_{AC}$ and $\mathcal{P}_{AB}$ are perpendicular to $\mathcal{P}$. Since the geometry in $\mathcal{S}$ is Euclidean, $b = c \cos \alpha$; hence, from (4.9-4),

$$b' = (c' \cos \alpha)(1 + \varepsilon_3). \tag{4.9-5}$$

We now repeat the entire procedure with the same triangle in $\mathcal{S}$, but this time we place the vertex B at the point Z of tangency. We find

$$a' = (c' \cos \beta)(1 + \varepsilon_4),$$

where β is the angle at B. As $c \to 0$, the angular defect of the triangle $\to 0$, hence $\cos \beta = (\sin \alpha)(1 + \varepsilon_5)$, hence

$$a' = (c' \sin \alpha)(1 + \varepsilon_6). \tag{4.9-6}$$

From (4.9-5) and (4.9-6) we find furthermore

$$a'^2 + b'^2 = c'^2(1 + \varepsilon_7). \tag{4.9-7}$$

Equations (4.9-5)–(4.9-7) give the desired fine structure for $\mathbb{H}^2$; since Z was an arbitrary point of $\mathcal{P}$, those equations hold throughout the plane.

Chapter 5

Differential Geometry of Surfaces

This chapter is intended to help the reader learn some of the concepts of differential geometry as they appear in terms of coordinates and also to move toward thinking of them in intrinsic terms ('coordinate free'), which is the modern approach to differential geometry. Our primary interest is in the hyperbolic plane as a Riemannian manifold.

In the calculus of several variables surfaces in $\mathbb{R}^3$ are important objects of study. The main technique for working with these surfaces is to parametrize them or parts of them, as one does with curves. The parameters can be thought of as coordinates in the part of the surface that is parametrized, because the points there are in one-to-one correspondence with certain pairs of real numbers, namely those that are in the parameter set. Geometric objects and quantities are then computed in terms of the coordinates. Some examples of these are discussed in this chapter: tangent vectors, lengths of curves, angles between curves, parallel transport of vectors, the Riemannian metric (metric tensor), areas of regions on a surface, (Gaussian) curvature of surfaces, and isometries. In order for such a quantity as the curvature of a surface to be an intrinsic feature of the surface, it must be true that the results computed using any two parametrizations are the same. Sometimes the quantity can be defined without explicit reference to coordinates, and then it is called a "coordinate free" definition. However the definition is made, it is rare to be able to make a computation without using coordinates (parameters).

Once the definitions are properly formulated they can be used in a more general and abstract setting. We can go to higher dimensions and also to 'spaces' called *differentiable manifolds.* This notion has been found to be very useful, justifying the extra level of abstraction. The main point in this chapter is to get the freedom of working with surfaces that are not necessarily embedded in $\mathbb{E}^3$, because the hyperbolic plane cannot be isometrically embedded into $\mathbb{E}^3$. Manifolds will be treated more thoroughly in Chapter 9, but this chapter lays the groundwork by studying surfaces in order to see how to formulate the definitions. Calculations are made strictly in terms of coordinates until Sections 10 and 11, which show how to define tangent vectors, tensors, vector fields and tensor fields intrinsically.

5.1 Parametric Representation of a Surface in Three Dimensions

To describe a surface or piece of a surface in a three-dimensional space, we let Ω be a convex region (open set) in a parameter space, with coordinates u and v, as in Fig. 5.1a. Convexity means that Ω contains the segment joining any two of its points. Thus Ω could be a disk but not an annulus. Then, a surface S, indicated schematically in Fig. 5.1b, is given by the equations

$$\left.\begin{array}{l} X = X(u,v) \\ Y = Y(u,v) \\ Z = Z(u,v) \end{array}\right\} \quad u,v \in \Omega, \tag{5.1-1}$$

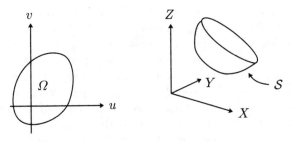

Fig. 5.1a   **Fig. 5.1b**

where X, Y, Z are Cartesian coordinates in 3-space. We assume that the functions $X(u,v), Y(u,v)$, and $Z(u,v)$ are smooth (at least C^2, which means that every derivative of order at most two exists and is continuous) and are such that the above equations determine a one-to-one relation between the points of Ω and the resulting points (X, Y, Z) which constitute the surface S. The parameters u and v are called *coordinates* or *curvilinear coordinates* on the surface S. Examples are the following:

Sphere:

$$\left.\begin{array}{l} X = a \sin\phi \cos\theta \\ Y = a \sin\phi \sin\theta \\ Z = a \cos\phi \end{array}\right\} \quad \begin{array}{l} \Omega: 0 < \phi < \pi,\ 0 < \theta < 2\pi \\ a = \text{const} > 0 \end{array} \tag{5.1-2}$$

Torus:

$$\left.\begin{array}{l} X = (a + b\cos\alpha)\cos\theta \\ Y = (a + b\cos\alpha)\sin\theta \\ Z = b\sin\alpha \end{array}\right\} \quad \begin{array}{l} \Omega: 0 < \alpha < 2\pi,\ 0 < \theta < 2\pi \\ a, b\ \text{const},\ a > b > 0 \end{array} \tag{5.1-3}$$

Saddle:

$$\left.\begin{array}{l} X = u \\ Y = v \\ Z = uv. \end{array}\right\} \quad \Omega: \text{ all } u, \text{ all } v.$$

Comments: (1) In the first two examples, it was necessary to omit certain points (in the case of the sphere, the poles and the prime meridian), in order to have a one-to-one smooth relation between Ω and S; generally, for a closed surface like a complete sphere or a complete torus, one must have two or more overlapping coordinate systems. To cover the sphere, one might have one coordinate system that covers slightly more than the northern hemisphere and one that covers slightly more than the southern hemisphere, the two systems overlapping in a band around the equator. We shall require that if u, v and u', v' are the coordinates of the two systems, then, in the part of the surface where they overlap, u and v are single-valued twice differentiable functions of u' and v', and conversely. The part of the surface covered by one coordinate system, with the coordinates of that system defined in it, is called a *coordinate chart*. Overlapping coordinate charts will appear again in Section 9.1 in connection with the definition of *manifold*. (2) Many mathematicians object to using a single symbol (like X or Y or Z) to denote both a variable and a function, as was done in (5.1-1). We take the attitude that in many applications (especially in physics) there are not enough letters for everything and one must be able to tell from the context whether a particular symbol denotes a dependent or independent quantity.

Embeddings. For some purposes, it is convenient to assume that the mapping from u, v to X, Y, Z is not only continuous, but has, in a certain sense, a continuous inverse. If u_0, v_0 is a given point of Ω, and X_0, Y_0, Z_0 is its image under the mapping (5.1-1), continuity at u_0, v_0 means that for any positive ϵ there is a positive δ such that

$$\text{If } \sqrt{(u - u_0)^2 + (v - v_0)^2} < \delta, \tag{5.1-4}$$

$$\text{then } \sqrt{(X - X_0)^2 + (Y - Y_0)^2 + (Z - Z_0)^2} < \epsilon, \tag{5.1-5}$$

where $X = X(u, v)$, and so on. By the continuity of the inverse, we mean that for any positive δ there is a positive ϵ such that if (5.1-5) holds, for any X, Y, Z on the surface, then the preimage u, v of X, Y, Z satisfies (5.1-4). When this additional condition is satisfied, we say that the surface S is *embedded* in the three-dimensional space.

Then, the topology that the surface S inherits from the 3-space is the same as the topology of the plane region Ω, which is simply the topology of an open disk. If the condition is not satisfied, as in Exercise 5 below, it is possible to find a sequence of points in the surface which converges in the 3-space to X_0, Y_0, Z_0, even though the preimages do not converge to u_0, v_0. In all our examples (except that in Exercise 5), the embedding condition is satisfied. From Section 5.3 on, we shall be concerned with abstract surfaces, which are not assumed to lie in any other space; hence the question of embedding does not arise.

Exercises

1. Find the functions $X(u,v)$, $Y(u,v)$, $Z(u,v)$ for a stereographic projection of the u, v-plane onto a sphere of unit radius (minus its north pole) resting on the plane with its south pole at the origin $u = v = 0$. The projection is obtained by drawing a straight line from the north pole N to a point u, v of the plane, as indicated in Fig. 5.1c; the point X, Y, Z is the intersection of that line with the sphere.

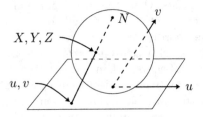

Fig. 5.1c.

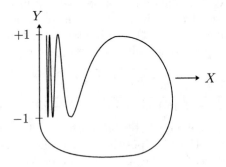

Fig. 5.1d.

2. Give a proof (or informal argument) that it is not possible to cover an entire sphere with a single coordinate system.
3. What is the smallest possible number of coordinate systems that will cover a torus?

4. Which points of the torus are omitted by the coordinates α, θ described above? Which points would be omitted if the plus sign were replaced by a minus sign in the first two equations of (5.1-3)?

5. A part of the curve in Fig. 5.1d is the graph of the equation $Y = \sin(1/X)$ for $0 < X < 1$; another part lies on the Y-axis; assume that this curve is the cross section of a surface obtained by letting Z vary. Show that the embedding condition is not satisfied, as stated in the text.

5.2 Lengths of Curves; the Line Element; the Metric Tensor

Suppose that (u, v) and $(u + \Delta u, v + \Delta v)$ are two points in Ω quite close together. Then the corresponding points (X, Y, Z) and $(X + \Delta X, Y + \Delta Y, Z + \Delta Z)$ of S are also quite close together, and the distance between them is $\Delta s = \sqrt{\Delta X^2 + \Delta Y^2 + \Delta Z^2}$, where, to first order of small quantities,

$$\Delta X = \frac{\partial X}{\partial u} \Delta u + \frac{\partial X}{\partial v} \Delta v,$$

$$\Delta Y, \ \Delta Z \text{ similar};$$

hence, we have

$$\Delta s^2 \approx g_{11} \Delta u^2 + 2 g_{12} \Delta u \Delta v + g_{22} \Delta v^2 \tag{5.2-1}$$

where

$$g_{11} = g_{11}(u, v) = \left(\frac{\partial X}{\partial u}\right)^2 + \left(\frac{\partial Y}{\partial u}\right)^2 + \left(\frac{\partial Z}{\partial u}\right)^2$$

$$g_{12} = g_{12}(u, v) = \frac{\partial X}{\partial u}\frac{\partial X}{\partial v} + \frac{\partial Y}{\partial u}\frac{\partial Y}{\partial v} + \frac{\partial Z}{\partial u}\frac{\partial Z}{\partial v} \tag{5.2-2}$$

$$g_{22} = g_{22}(u, v) = \left(\frac{\partial X}{\partial v}\right)^2 + \left(\frac{\partial Y}{\partial v}\right)^2 + \left(\frac{\partial Z}{\partial v}\right)^2$$

It is customary to paraphrase (5.2-1) as

$$ds^2 = g_{11}du^2 + 2g_{12}dudv + g_{22}dv^2. \tag{5.2-3}$$

The expression ds is called the *line element* and the above equation (when the g_{jk} are specified functions of u, v) is the *equation of the line element* in the surface S.

Now let Γ be a curve in Ω and C the corresponding curve in S, and let t be a parameter on Γ, so that Γ is represented by functions $u(t), v(t)$, for $a \le t \le b$. Then, the curve C on the surface is represented by the functions

$X(u(t), v(t))$, etc. We wish to find the length of C. We approximate C by a polygonal curve by partitioning the interval $[a, b]$ as

$$a = t_0 < t_1 < \cdots < t_k = b. \tag{5.2-4}$$

For each increment $\Delta t = t_j - t_{j-1}$, the parameters u, v have increments given approximately by

$$\Delta u = \dot{u} \Delta t, \qquad \Delta v = \dot{v} \Delta t,$$

where the dot denotes differentiation of the functions $u(t)$ and $v(t)$. The length of the corresponding segment of the curve C in S is approximately

$$\Delta s \approx \left[g_{11} \Delta u^2 + 2g_{12} \Delta u \Delta v + g_{22} \Delta v^2 \right]^{1/2}$$

$$\approx \left[g_{11} \dot{u}^2 + 2g_{12} \dot{u}\dot{v} + g_{22} \dot{v}^2 \right]^{1/2} \Delta t$$

By adding the lengths of all such segments and taking the limit as the partition (5.2-4) is refined, to give a Riemann integral, we find that the *length* of the curve C is

$$\text{length } (C) = \int_a^b \left[g_{11} \dot{u}^2 + 2g_{12} \dot{u}\dot{v} + g_{22} \dot{v}^2 \right]^{1/2} dt, \tag{5.2-5}$$

where g_{11} is understood as $g_{11}(u(t), v(t))$, and so on.

When we write $g_{21} = g_{12}$ and arrange the four quantities g_{jk} as a 2×2 (symmetric) matrix, we have what is called the (*Riemannian*) *metric tensor of the surface S with respect to the coordinates u, v:*

$$G = \begin{pmatrix} g_{11} & g_{12} \\ g_{21} & g_{22} \end{pmatrix}. \tag{5.2-6}$$

The matrix G varies from one point of the surface to another, that is, is a function of u and v; furthermore, it depends on the particular choice of the curvilinear coordinates u, v on the surface, as indicated by (5.2-2). For example, if the surface S is a plane and u, v are Cartesian coordinates x, y, the line element and metric tensor are

$$ds^2 = dx^2 + dy^2, \qquad G = \begin{pmatrix} 1 & 0 \\ 0 & 1 \end{pmatrix},$$

while if u, v are polar coordinates r, θ, we have

$$ds^2 = dr^2 + r^2 d\theta^2, \qquad G = \begin{pmatrix} 1 & 0 \\ 0 & r^2 \end{pmatrix}.$$

It will be seen in Section 5.4 below that the entire intrinsic geometry of S can be obtained from the metric tensor (or equivalently from the line element). For example, the geodesics are the analogues of straight lines in

Euclidean geometry. If $\mathcal{C}$ is a curve from P to Q and is the shortest curve from P to Q, then it is a *geodesic*, and it will be seen that the geodesics in a surface can be determined by certain differential equations in which the functions $g_{jk}(u,v)$ appear. The angle of intersection of two curves and the area bounded by a closed curve can also be determined in terms of the $g_{jk}(u,v)$.

It is recalled from linear algebra that a (real symmetric) matrix $\mathcal{M}$ is *positive definite* if $\vec{v}^T M \vec{v} > 0$ for all nonzero vectors $\vec{v}$. The matrix G is positive definite, because, first, the quantity $\Delta s^2 = \Delta X^2 + \Delta Y^2 + \Delta Z^2$ is positive, and then Equation (5.2-1) can be written as $\Delta s^2 = \vec{v}^T G \vec{v}$, where $\vec{v}$ is the vector

$$\vec{v} = \begin{pmatrix} \Delta u \\ \Delta v \end{pmatrix}.$$

To simplify the integrand of (5.2-5), we define a tangent vector to Γ

$$\vec{w} = \vec{w}(t) = \begin{pmatrix} \dot{u} \\ \dot{v} \end{pmatrix},$$

and then the quantity in the square brackets in equation (5.2-5) is $\vec{w}^T G \vec{w}$, and we have

$$\text{length } (\mathcal{S}) = \int_a^b [\vec{w}^T G \vec{w}]^{1/2} dt. \tag{5.2-7}$$

The choice of the curvilinear coordinates u, v in a given surface $\mathcal{S}$ is highly arbitrary. In the Euclidean plane, one thinks of Cartesian coordinates as being in a sense the primary choice, and others, such as polar coordinates or elliptic coordinates, as being secondary. In a general surface, there are no Cartesian coordinates; hence, we must put all systems on equal footing, and we must ensure that geometrical concepts like lengths of curves and angles between curves are so defined as to be independent of the choice of the system.

Suppose that u, v is one choice, and u', v' is another. Then, u, v are functions (assumed at least once continuously differentiable) of u', v' and we write

$$u = \hat{u}(u', v'), \qquad v = \hat{v}(u', v').$$

We denote by (g_{jk}) and (g'_{jk}) the metric tensors for the two systems. From (5.2-2) by the chain rule for differentiation, we find

$$g'_{11} = g_{11} \left(\frac{\partial \hat{u}}{\partial u'} \right)^2 + 2g_{12} \frac{\partial \hat{u}}{\partial u'} \frac{\partial \hat{v}}{\partial u'} + g_{22} \left(\frac{\partial \hat{v}}{\partial u'} \right)^2$$

$$g'_{12} = g_{11} \frac{\partial \hat{u}}{\partial u'} \frac{\partial \hat{u}}{\partial v'} + g_{12} \left(\frac{\partial \hat{u}}{\partial u'} \frac{\partial \hat{v}}{\partial v'} + \frac{\partial \hat{u}}{\partial v'} \frac{\partial \hat{v}}{\partial u'} \right) + g_{22} \frac{\partial \hat{v}}{\partial u'} \frac{\partial \hat{v}}{\partial v'} \tag{5.2-8}$$

$$g'_{22} = g_{11} \left(\frac{\partial \hat{u}}{\partial v'} \right)^2 + 2g_{12} \frac{\partial \hat{u}}{\partial v'} \frac{\partial \hat{v}}{\partial v'} + g_{22} \left(\frac{\partial \hat{v}}{\partial u'} \frac{\partial \hat{v}}{\partial v'} \right)^2$$

Note: Certain terms have been combined by taking advantage of the relation $g_{12} = g_{21}$.

If we introduce the Jacobian matrix J by

$$J = \begin{pmatrix} j_{11} & j_{12} \\ j_{21} & j_{22} \end{pmatrix} = \begin{pmatrix} \dfrac{\partial \hat{u}}{\partial u'} & \dfrac{\partial \hat{u}}{\partial v'} \\ \dfrac{\partial \hat{v}}{\partial u'} & \dfrac{\partial \hat{v}}{\partial v'} \end{pmatrix}, \tag{5.2-9}$$

then equations (5.2-8) can be written as

$$g'_{ik} = \sum_{r,s=1}^{2} j_{ri} g_{rs} j_{sk}.$$

In matrix notation, this is just the product

$$G' = J^T G J. \tag{5.2-10}$$

This is the *transformation law* for the matrix of the metric tensor.

In differential geometry, it is customary to omit the circumflex accent in equations like (5.2-8) and (5.2-9). To make that possible, we adopt the *convention* that when a derivative $(\partial u)/(\partial u')$ is encountered, it is understood that u is a function of two variables (or more, in higher-dimensional cases). To find out what those variables are, one looks at the denominator. If a primed variable appears there, the independent variables are understood to be u' and v'; hence, in this example, v' is held constant during the partial differentiation. With that convention, the Jacobian matrix of the inverse transformation is

$$K = \begin{pmatrix} \dfrac{\partial u'}{\partial u} & \dfrac{\partial u'}{\partial v} \\ \dfrac{\partial v'}{\partial u} & \dfrac{\partial v'}{\partial v} \end{pmatrix}.$$

From the chain rule for differentiation, we see that the matrix product KJ is the Jacobian matrix of the identity transformation from u, v to u, v. Hence $KJ = \begin{pmatrix} 1 & 0 \\ 0 & 1 \end{pmatrix}$; that is, $K = J^{-1}$.

We can now see directly that the length of a curve is invariant under transformations of curvilinear coordinates. Namely, by the chain rule for differentiation of $u'(u(t), v(t))$ with respect to t, we have

$$\vec{w}' = \begin{pmatrix} \dot{u}' \\ \dot{v}' \end{pmatrix} = K \begin{pmatrix} \dot{u} \\ \dot{v} \end{pmatrix}.$$

Hence,

$$\vec{w}'^T G' \vec{w}' = \vec{w}^T K^T J^T G J K \vec{w} = \vec{w}^T G \vec{w}, \tag{5.2-11}$$

so that

$$\int_a^b [\vec{w}'^T G' \vec{w}']^{1/2} dt = \int_a^b [\vec{w}^T G \vec{w}]^{1/2} dt,$$

as required. In connection with the next section, we note that this proof of
the invariance of length (C) makes no use of the original coordinates X, Y, Z,
hence is independent of how or whether the surface S can be embedded in
Euclidean 3-space. When the surface *can* be embedded, as described in this
section, and its metric tensor is defined by (5.2-2), we say that its geometry
(or its metric) is *inherited* from the geometry (or metric) of the Euclidean
3-space in which it is embedded.

Exercises

1. Find the metric tensor for the sphere, the torus, and the saddle, in
    terms of the coordinates given in the preceding section and for the
    sphere in terms of the coordinates u, v of Exercise 1 in that section.
2. Suppose the curve Γ in the u, v-plane is simple and closed, that is, is
    nonself-intersecting except that $u(a) = u(b)$ and $v(a) = v(b)$. Is the
    curve C on S also simple and closed?
3. Show how the discussion can be modified (perhaps informally) to in-
    clude simple closed curves that link with the torus by going round
    the Z-axis or round the axis of the torus. (These curves pass through
    the points on the torus that were left undefined by the representation
    (5.1-3). To investigate whether a simple closed curve can go arbitrarily
    many times round the Z-axis and arbitrarily many times round the
    axis of the torus, consider the curve given in terms of a parameter t by
    the equations

$$\alpha = pt \pmod{2\pi}, \quad \theta = qt \pmod{2\pi},$$

    where α and θ are as in the representation (5.1-3) and where p and q
    are positive integers. What relation between p and q is necessary for
    the resulting curve to be a *simple* closed curve?

5.3 Abstract Geometric Surfaces; Line Element in the Hyperbolic Plane

We now take a more abstract point of view. We define a *surface geometry*
or *two-dimensional geometry* by specifying a region Ω of variation of the
parameters u, v and specifying four smooth functions $g_{jk}(u, v)$ (with $g_{12} = g_{21}$) for u, v in Ω, without saying anything about whether the surface even
can be embedded in 3-space, and without requiring that the metric tensor
$G = (g_{jk})$ be obtainable by formulas of the kind (5.2-2), but see the Lemma
below. We still require that under a transformation of the parameters the
metric tensor transform by (5.2-8). That has the consequence that certain

geometrical concepts are independent of the choice of the parameters u, v, for example, the length of a curve, as shown at the end of the preceding section, and the angle between two curves, as shown in Section 5.5 below. The matrix G must be positive definite to avoid negative distances.

An example is provided by the hyperbolic plane, with u, v taken as the coordinates ξ, η described in Section 3.9. According to that section, if two points are separated by an increment $\Delta\xi$ at constant η, the distance between them is equal to $\Delta\xi$; if they are separated by an increment $\Delta\eta$ at constant ξ, the distance is $e^\xi \Delta\eta + \cdots$ (where the dots represent terms of higher order due to the fact that the horocycle is approximately a straight line only for small distances). Those two displacements are orthogonal, because the lines of the asymptotic pencil (given by $\eta = $ const) are everywhere perpendicular to the horocycles (given by $\xi = $ const). Therefore, the approximate Pythagorean law (4.9-7) holds, and the distance Δs that results from the two displacements together is given by

$$\Delta s^2 = \Delta\xi^2 + e^{2\xi}\Delta\eta^2 + \cdots .$$

Therefore, taking the limit as in the preceding section gives the line element and the metric tensor as

$$ds^2 = d\xi^2 + e^{2\xi}d\eta^2, \qquad G = \begin{pmatrix} 1 & 0 \\ 0 & e^{2\xi} \end{pmatrix} \qquad (5.3\text{-}1)$$

If a curve $\mathcal{C}$ in the hyperbolic plane is represented by smooth functions $\xi(t)$ and $\eta(t)$, for $a \le t \le b$, then the length of $\mathcal{C}$ is given, as in (5.2-5), by

$$L(\mathcal{C}) = \int_a^b [\dot{\xi}^2 + e^{2\xi}\dot{\eta}^2]^{1/2}dt. \qquad (5.3\text{-}2)$$

The region Ω consists of all real ξ and all real η. Transformations from the coordinates ξ, η to other coordinates in the hyperbolic plane are considered in Chapter 7.

We note that the same argument that led to (5.3-1) could have been applied to polar coordinates. Displacements corresponding to increments Δr and $\Delta\theta$ are of magnitude Δr and $g(r)\Delta\theta$ and are perpendicular, hence

$$ds^2 = dr^2 + g(r)^2 d\theta^2. \qquad (5.3\text{-}3)$$

The difference is that while (5.3-1) is explicit, we don't yet have a formula for $g(r)$.

We now turn briefly to the problem of whether a surface that is given abstractly, as here, can be embedded in Euclidean 3-space $\mathbb{E}^3$ and its metric inherited from $\mathbb{E}^3$. We assume that the functions $g_{jk}(u, v)$ are given. Then the problem is to regard equations (5.2-2) as partial differential equations for the functions $X(u, v)$, $Y(u, v)$, and $Z(u, v)$. We state the main result without proof.

Lemma: If the matrix

$$G = \begin{pmatrix} g_{11}(u,v) & g_{12}(u,v) \\ g_{21}(u,v) & g_{22}(u,v) \end{pmatrix}$$

is symmetric and positive definite in the region Ω, if the g_{ij} are continuously differentiable functions of u, v, and if u_0, v_0 is a point of Ω, then there is a neighborhood of the point u_0, v_0 in which the Equations (5.2-2) have solutions such that the mapping $u, v \to X, Y, Z$ is a one-to-one relation between the points of that neighborhood and a piece of surface in $\mathbb{E}^3$.

We say that the abstract surface can be represented locally in the Euclidean $\mathbb{E}^3$. Clearly, the solution of the differential equations is not unique, because the piece of surface can be moved about in $\mathbb{E}^3$ by any rigid motion, and can in general be warped or bent to some extent, without changing its intrinsic geometry. The n-dimensional case is this: Consider an n-dimensional Riemannian space R^n with curvilinear coordinates $u^1, \ldots, u^n$ and metric tensor an $n \times n$ symmetric positive definite matrix, as will be discussed in Chapter 9. Then, R^n can be represented locally as an n-dimensional region in a Euclidean space $\mathbb{E}^m$ if $m = n(n+1)/2$. In the case considered above, $n = 2$ and $m = 3$.

In Chapter 7, a portion of the hyperbolic plane is embedded in this way in Euclidean 3-space. It is not possible to represent the entire hyperbolic plane as a surface in $\mathbb{E}^3$ without self-intersection and overlapping of the surface. However, the entire hyperbolic plane can be represented as a surface in three-dimensional Minkowski space; see Sections 9.6 and 10.2.

5.4 Geodesics and the Calculus of Variations

We denote the quantity in the square brackets in (5.2-5) by Φ; that is,

$$\Phi(u,v,\dot{u},\dot{v}) = g_{11}(u,v)\dot{u}^2 + 2g_{12}(u,v)\dot{u}\dot{v} + g_{22}(u,v)\dot{v}^2. \qquad (5.4\text{-}1)$$

A convenient choice of the parameter t on any curve is the arclength s, that is, the length of the portion of the curve from some fixed point to a variable point along the curve. Then, the quantity on the left side of (5.2-5) is equal to $b - a$. That is,

$$b - a = \int_a^b \left[\Phi\left(u(s), v(s), \frac{du}{ds}, \frac{dv}{ds}\right) \right]^{1/2} ds.$$

If we keep a fixed and let b vary, differentiation of this equation with respect to b gives

$$1 = \left[\Phi\left(u, v, \frac{du}{ds}, \frac{dv}{ds}\right) \right]^{1/2},$$

taken at $s = b$, but b was an arbitrary point. That is, if the parameter t is the arclength s, then the quantity Φ has the constant value 1 on the curve.

Now let C_0 be a curve from a point P on S to a point Q on S, determined by functions $u_0(s), v_0(s)$. We wish to find conditions such that C_0 has smaller length than any neighboring curve C from P to Q. Let C be determined by functions

$$u(s) = u_0(s) + \varepsilon f(s),$$
$$v(s) = v_0(s) + \varepsilon g(s),$$
(5.4-2)

where f and g are arbitrary smooth functions (continuously differentiable), which vanish at the points P and Q; that is,

$$f(a) = f(b) = 0, \qquad g(a) = g(b) = 0. \tag{5.4-3}$$

Note: We are assuming that s is arclength along C_0 but not along neighboring curves — in fact, it can't be, because neighboring curves have length greater than $b - a$.

We denote the length of C by $L(\varepsilon)$, so that

$$L(\varepsilon) = \int_a^b \left[\Phi\left(u, v, \frac{du}{ds}, \frac{dv}{ds}\right) \right]^{1/2} ds,$$

where $u(s)$ and $v(s)$ are as in (5.4-2). We require that, for any choice of the smooth functions $f(s), g(s)$ that vanish at the endpoints, $L(\varepsilon)$ have a minimum at $\varepsilon = 0$. Hence, we require that $(d/d\varepsilon) L(\varepsilon)$ be equal to 0 for $\varepsilon = 0$. Because of the smoothness of the functions involved, we can differentiate under the integral sign in the above equation; hence we require that

$$0 = \int_a^b \left[\frac{1}{2} \Phi^{-1/2} \frac{d}{d\varepsilon} \Phi \right]_{\varepsilon=0} ds.$$

The factor $\Phi^{-1/2}$ can be dropped, because it is equal to 1 for $\varepsilon = 0$. By the chain rule for differentiation,

$$\frac{d}{d\varepsilon} \Phi = \frac{\partial \Phi}{\partial u} f + \frac{\partial \Phi}{\partial v} g + \frac{\partial \Phi}{\partial \dot{u}} \frac{df}{ds} + \frac{\partial \Phi}{\partial \dot{v}} \frac{dg}{ds}.$$

This quantity has to be integrated from $s = a$ to $s = b$. The last two terms can be integrated by parts, for example,

$$\int_a^b \frac{\partial \Phi}{\partial \dot{u}} \frac{df}{ds} ds = - \int_a^b \left(\frac{d}{ds} \frac{\partial \Phi}{\partial \dot{u}} \right) f(s) ds.$$

[The "integrated part" $(\partial \Phi/\partial \dot{u}) f|_a^b$ vanishes because of (5.4-3).] Hence, the requirement is that

$$0 = \int_a^b \left[\frac{\partial \Phi}{\partial u} - \frac{d}{ds} \frac{\partial \Phi}{\partial \dot{u}} \right] f(s) ds + \int_a^b \left[\frac{\partial \Phi}{\partial v} - \frac{d}{ds} \frac{\partial \Phi}{\partial \dot{v}} \right] g(s) ds,$$

for $\varepsilon = 0$ in the bracketed quantities. Because of the arbitrary nature of $f(s)$ and $g(s)$, this requires that along C_0

$$\frac{\partial \Phi}{\partial u} - \frac{d}{ds}\frac{\partial \Phi}{\partial \dot u} = 0, \qquad \frac{\partial \Phi}{\partial v} - \frac{d}{ds}\frac{\partial \Phi}{\partial \dot v} = 0 \quad (a \le s \le b). \tag{5.4-4}$$

In these equations, it is understood that $u(s) = u_0(s)$ and $v(s) = v_0(s)$, because $\varepsilon = 0$. Then, these equations ensure that $(d/d\varepsilon)\, L(\varepsilon) = 0$ for $\varepsilon = 0$ for all choices of $f(s)$ and $g(s)$. They are the *equations of a geodesic*, and the curve C_0 is the *geodesic* from P to Q.

[More explicit differential equations for a geodesic are given in (5.6-6) below.]

Example: Consider the line element of the hyperbolic plane in coordinates ξ, η given, according to (5.3-1), by $ds^2 = d\xi^2 + e^{2\xi} d\eta^2$, where we assume the unit of length to be so chosen that the constant K in (3.9-3) is 1.

The function Φ is

$$\Phi = \Phi(\xi, \eta, \dot\xi, \dot\eta) = \dot\xi^2 + e^{2\xi}\dot\eta^2. \tag{5.4-5}$$

Equations (5.4-4) of a geodesic are, after dropping a factor 2 from each,

$$e^{2\xi}\dot\eta^2 - \ddot\xi = 0, \qquad 0 - \frac{d}{ds}\left(e^{2\xi}\dot\eta\right) = 0.$$

The second of these gives

$$e^{2\xi}\dot\eta = \text{const} = A, \tag{5.4-6}$$

and substitution into the first gives

$$A^2 e^{-2\xi} - \ddot\xi = 0.$$

Multiplication of this equation through by $-2\dot\xi$ and integrating gives

$$A^2 e^{-2\xi} + \dot\xi^2 = B, \tag{5.4-7}$$

where B is another constant. Since Φ has the constant value 1 along the geodesic, we have, from Equations (5.4-5)-(5.4-7),

$$1 = \Phi = B - A^2 e^{-2\xi} + e^{2\xi}(Ae^{-2\xi})^2 = B.$$

For the case $A = 0$, see Equation (5.4-10) below. If $A \ne 0$, we call $\xi_0 = \log A$. With $B = 1$, (5.4-7) then gives

$$\dot\xi = \sqrt{1 - e^{-2(\xi - \xi_0)}}.$$

If we call $\zeta = e^{\xi - \xi_0}$ and multiply this last equation through by $e^{\xi - \xi_0}$, we have

$$\dot\zeta = \sqrt{\zeta^2 - 1},$$

which can be integrated by regarding s as a function of ζ so that

$$s = \int (\zeta^2 - 1)^{-1/2} d\zeta.$$

With the aid of a table of integrals, we find

$$s = \ln \left[\zeta + \sqrt{\zeta^2 - 1} \right] + s_0.$$

Since the origin of arclength along the geodesic is arbitrary, we can take $s_0 = 0$. Solving this equation for ζ then gives

$$\zeta = e^{\xi - \xi_0} = \cosh s \tag{5.4-8}$$

Then (5.4-6) gives

$$\dot{\eta} = A e^{-2\xi} = e^{-\xi_0} \frac{1}{\cosh^2 s};$$

on integrating, this gives $\eta = e^{-\xi_0} \tanh s + \eta_0$. The conclusion is that for the general straight line in the hyperbolic plane, ξ and η are functions of s given by

$$\left\{ \begin{array}{l} e^{\xi - \xi_0} = \cosh s \\ \eta - \eta_0 = e^{-\xi_0} \tanh s \end{array} \right\}, \tag{5.4-9}$$

where ξ_0 and η_0 are arbitrary constants. If $A = 0$, we have $\eta = \text{const}$ from (5.4-6) and $\dot{\xi} = 1$ from (5.4-7); hence we can take $\xi = s$, and we have

$$\left\{ \begin{array}{l} \xi = s \\ \eta = \eta_0 \end{array} \right\}, \tag{5.4-10}$$

which are simply the equations of the lines of the asymptotic bundle that was used to define the coordinates.

Note on Nonminimizing Geodesics: Any path $u(s), v(s)$ that satisfies Equations (5.4-4) is called a *geodesic*, even though it may not be the shortest path between the endpoints P and Q. For example, consider spherical geometry, and let C_0 be one-half of the equator of a sphere; then the points P and Q are antipodal, and any other great circle route from P to Q has the same length as C_0. Even worse, let C_0 be a portion of the equator covering more than 180°, as in Fig. 5.4. Then, C_0 is a geodesic, but there are curves C (even ones close to C_0) with a smaller length. That cannot happen in hyperbolic geometry; see Exercise 1 below. Furthermore, on any surface S, according to a theorem of J. H. C. Whitehead, if we restrict consideration to small enough regions (for example, on the sphere, spherical caps smaller than a hemisphere), then all geodesics confined to such a region actually minimize the length of curves between the endpoints P and Q and lying within the region. The theorem of Whitehead appears as Theorem 6.2 in

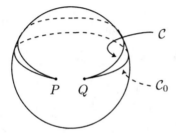

Fig. 5.4.

Chapter I of a book by S. Helgason, *Differential Geometry and Symmetric Spaces,* Academic Press, New York and London, 1962.

Exercise

1. Show that in hyperbolic geometry, a curve $\mathcal{C}$ from P to Q is never shorter than the geodesic $\mathcal{C}_0$ (i.e., the straight line segment) from P to Q. *Hint:* Approximate $\mathcal{C}$ by a polygonal curve and use the triangle inequality (Axiom 2).

5.5 Angles

Let the functions $X(t), Y(t), Z(t)$ determine a curve $\mathcal{C}$ in space. The vector $\vec{V}$ whose components are $\dot{X}, \dot{Y}, \dot{Z}$ is a *tangent vector* to $\mathcal{C}$. Its magnitude or length is

$$\|\vec{V}\| = \left[\dot{X}^2 + \dot{Y}^2 + \dot{Z}^2 \right]^{1/2}. \tag{5.5-1}$$

If $X'(t), Y'(t), Z'(t)$ (where the prime does *not* indicate differentiation) is another curve $\mathcal{C}'$, and if the two curves intersect at some point P_0, say at $t = t_0$, then, according to standard linear algebra, the angle θ between the two tangent vectors is determined by the equation

$$\vec{V} \cdot \vec{V}' = \|\vec{V}\| \|\vec{V}'\| \cos\theta. \tag{5.5-2}$$

Now suppose that the curves $\mathcal{C}$ and $\mathcal{C}'$ lie on a surface $\mathcal{S}$, in which curvilinear coordinates u, v have been defined and the metric tensor has been determined according to (5.2-2). We wish to express the magnitudes and angles of vectors in terms of the quantities $u, v, g_{jk}(u, v)$. The curves $\mathcal{C}, \mathcal{C}'$ are now given by functions $u(t), v(t)$ and $u'(t), v'(t)$ (where, again, the prime does not indicate differentiation), and the coordinates X, Y, Z, as functions on the curves, are given by $X(u(t), v(t))$, and so on. By the chain rule of differentiation,

$$\dot{X} = \frac{\partial X}{\partial u}\,\dot{u} + \frac{\partial X}{\partial v}\,\dot{v}; \qquad \dot{Y},\,\dot{Z}\ \text{similar}.$$

Hence, the inner product $\vec{V}\cdot\vec{V}'$ is given by

$$
\begin{aligned}
\vec{V}\cdot\vec{V}' &= \dot{X}\dot{X}' + \dot{Y}\dot{Y}' + \dot{Z}\dot{Z}' \\
&= \left(\frac{\partial X}{\partial u}\,\dot{u} + \frac{\partial X}{\partial v}\,\dot{v}\right)\left(\frac{\partial X}{\partial u}\,\dot{u}' + \frac{\partial X}{\partial v}\,\dot{v}'\right) + \cdots \\
&= \left(\frac{\partial X}{\partial u}\right)^2 \dot{u}\dot{u}' + \frac{\partial X}{\partial u}\frac{\partial X}{\partial v}(\dot{u}\dot{v}' + \dot{v}\dot{u}') + \left(\frac{\partial X}{\partial v}\right)^2 \dot{v}\dot{v}'
\end{aligned}
$$

$$+ \cdots \ (\text{terms involving } Y \text{ and terms involving } Z).$$

When the indicated summations are performed, the net coefficient of $\dot{u}\dot{u}'$ is equal to $g_{11}(u,v)$, according to the definition (5.2-2). The other terms are similar, and we have

$$\vec{V}\cdot\vec{V}' = g_{11}\dot{u}\dot{u}' + g_{12}(\dot{u}\dot{v}' + \dot{v}\dot{u}') + g_{22}\dot{v}\dot{v}'.$$

We therefore define two-component vectors

$$\vec{w} = \begin{pmatrix} \dot{u} \\ \dot{v} \end{pmatrix}, \qquad \vec{w}' = \begin{pmatrix} \dot{u}' \\ \dot{v}' \end{pmatrix}, \tag{5.5-3}$$

and we define an inner product for such vectors at the point P_0 of the surface by

$$(\vec{w},\vec{w}') \overset{\text{def}}{=} g_{11}\dot{u}\dot{u}' + g_{12}(\dot{u}\dot{v}' + \dot{v}\dot{u}') + g_{22}\dot{v}\dot{v}', \tag{5.5-4}$$

where the $g_{jk}(u,v)$ are to be evaluated at P_0. We note in passing that this can also be written in vector-matrix notation as

$$(\dot{u}\ \dot{v})\begin{pmatrix} g_{11} & g_{12} \\ g_{21} & g_{22} \end{pmatrix}\begin{pmatrix} \dot{u}' \\ \dot{v}' \end{pmatrix} = \vec{w}^{T}G\vec{w}'. \tag{5.5-5}$$

The transformation from three-vector notation to two-vector notation is made according to the equation $\vec{V}\cdot\vec{V}' = (\vec{w},\vec{w}')$. Finally, then, the angle θ between two curves in the surface, at their point of intersection, is given, according to (5.5-2), by the equation

$$(\vec{w},\vec{w}') = (\vec{w},\vec{w})^{\frac{1}{2}}(\vec{w}',\vec{w}')^{\frac{1}{2}}\cos\theta. \tag{5.5-6}$$

If S is an abstract surface on which a line element or metric tensor has been defined, without reference to X,Y,Z, then the above equation is taken as *defining* the angle θ between two smooth curves at their point of intersection. In either case, for a given surface S, the inner products and the angle θ are invariant under transformations from one set of curvilinear coordinates to another, say u,v to u',v', when the metric tensor is transformed according to (5.2-8). (Here, the prime denotes a different coordinate system, not a second curve.)

Since geometrical concepts can be expressed in terms of points, lines (geodesics), lengths (distances), and angles, we now see that the geometry of a surface is completely determined by its line element or metric tensor.

If the parameter t on the curves is arclength, then $\|\vec{V}\| = \|\vec{V}'\| = 1$; then the tangent vectors are unit vectors, and the angle θ is given simply by

$$\cos\theta = \vec{V} \cdot \vec{V}' = (\vec{w}, \vec{w}'). \qquad (5.5\text{-}7)$$

We wish to show now that, even if S_0 is an abstract surface, as discussed in Section 5.3 above, the angle θ defined in this way is additive, in this sense:

Lemma: Let $\vec{w}$ and $\vec{w}'$ be unit vectors on the same side of the line containing the unit vector $\vec{w}_0$, as shown in Fig. 5.5, where these vectors are all tangent vectors at a point P_0 of the surface (i.e., are tangent vectors at P_0 to various smooth curves, perhaps geodesics, passing though P_0). Let $(\cdot, \cdot)$ denote the inner product (5.5-4) defined in terms of the metric tensor at P_0. Let θ, θ' be the angles between $\vec{w}_0$ and $\vec{w}, \vec{w}'$, respectively, and let α be the angle between $\vec{w}$ and $\vec{w}'$. Then $\alpha = |\theta - \theta'|$.

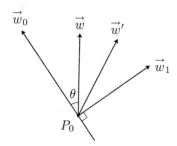

Fig. 5.5.

For the proof, see Exercises 1 and 2, below.

Since, furthermore, as $\vec{w}$ varies for given $\vec{w}_0$, θ varies from 0 to π, and since $\theta = 90°$ if $\vec{w} \perp \vec{w}_0$, we see that the angle θ defined here agrees with the angle determined by Axiom 5 for the case in which the abstract surface represents the hyperbolic plane.

Exercises

1. (Projection Formula) Let $\vec{w}_1$ be a unit vector orthogonal to $\vec{w}_0$ and on the same side of $\vec{w}_0$ as $\vec{w}$ and $\vec{w}'$, as in Fig. 5.5. Show that $(\vec{w}, \vec{w}_1) = \sin\theta$, and so on and that

$$\vec{w} = \cos\theta\,\vec{w}_0 + \sin\theta\,\vec{w}_1, \text{ etc.}$$

2. Prove the Lemma.
3. Show that if the metric tensor is diagonal, then the coordinate system is orthogonal.
4. Show that if the metric tensor is a multiple of the unit matrix, then the coordinate system is conformal.

5.6 Parallel Transport of Vectors

In the Euclidean plane, if we are given any two points P and Q, then vectors $\overrightarrow{PX}$ and $\overrightarrow{QY}$ originating at those points are parallel if and only if they make equal angles with the line through P and Q, in the manner indicated in Fig. 5.6.

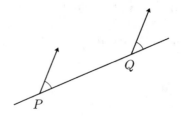

Fig. 5.6.

If we try to generalize that idea by saying that PX and QY are parallel if they make the same angles with respect to a *geodesic* from P to Q, we run into the following difficulty: Suppose that P, Q, and R are three points *not* on a common geodesic. If PX and QY make the same angle with respect to a geodesic through P and Q, and if QY and RZ make the same angle with respect to a geodesic through Q and R, then PX and RZ don't necessarily make the same angle with respect to a geodesic through P and R. That is, we should lose the transitivity of the notion of parallelism. We are therefore restricted to a notion of parallelism with respect to transport of a vector from P to Q along a specified curve. A rough physical analogy is this: consider a ship following a polygonal path on the earth's surface from P to Q. At P an arrow is placed on the ship's deck pointing in a specified direction. At each vertex of the polygonal path, when the ship suddenly changes bearing by turning through a certain angle, the arrow is turned through the same angle on the ship's deck, in the opposite direction, so as to maintain the same absolute "direction." On reaching Q, the arrow is said to have been carried there by "parallel transport." By consideration of special cases, one sees that if the thing is repeated, starting with the same vector at P but following a different polygonal path to Q, in general a different direction is obtained at point Q, because of the sphericity of the earth's surface.

Nevertheless, the general notion of parallel transport will be obtained by first considering parallel transport along a single geodesic.

We first put the equation (5.4-4) of a geodesic in a more convenient form. We rename the coordinates u, v as u^1, u^2 (where the superscripts are *not* exponents). Then the function Φ becomes

$$\Phi = \Phi(u^1, u^2, \dot{u}^1, \dot{u}^2) = \sum_{(j,k)} g_{jk}\dot{u}^j\dot{u}^k, \tag{5.6-1}$$

where $\sum_{(j,k)}$ is an abbreviation for $\sum_{j=1}^{2}\sum_{k=1}^{2}$, and the equations of a geodesic are

$$\frac{\partial\Phi}{\partial u^i} - \frac{d}{ds}\frac{\partial\Phi}{\partial \dot{u}^i} = 0 \qquad (i = 1, 2),$$

or explicitly

$$\sum_{(j,k)} \frac{\partial g_{jk}}{\partial u^i}\dot{u}^j\dot{u}^k - \frac{d}{ds}\left[\sum_{(k)}g_{ik}\dot{u}^k + \sum_{(j)}g_{ji}\dot{u}^j\right] = 0 \; (i = 1, 2) \tag{5.6-2}$$

When the differentiation of the square bracket is worked out, these equations contain $\ddot{u}^1$ and $\ddot{u}^2$; since they contain those quantities linearly, they can be solved. To do that, we denote by g^{rs} the entries in the reciprocal of the matrix G; that is,

$$\begin{pmatrix} g^{11} & g^{12} \\ g^{21} & g^{22} \end{pmatrix} = G^{-1} = \begin{pmatrix} g_{11} & g_{12} \\ g_{21} & g_{22} \end{pmatrix}^{-1}. \tag{5.6-3}$$

Note that these are both symmetric matrices. We multiply Equation (5.6-2) through by g^{im} and sum for $i = 1, 2$. Since $\sum_{(i)}g_{ik}g^{im} = \delta_k^m$ and $\sum_{(i)}g_{ji}g^{im} = \delta_j^m$, we find, after suitably renaming the summation indices in two of the terms, and changing the sign throughout

$$\sum_{(i)}g^{im}\sum_{(j,k)}\left[-\frac{\partial g_{jk}}{\partial u^i} + \frac{\partial g_{ik}}{\partial u^j} + \frac{\partial g_{ji}}{\partial u^k}\right]\dot{u}^j\dot{u}^k + 2\ddot{u}^m = 0.$$

We define the *Christoffel three-index symbol of the first kind* as

$$[jk, i] \stackrel{\text{def}}{=} \frac{1}{2}\left[-\frac{\partial g_{jk}}{\partial u^i} + \frac{\partial g_{ik}}{\partial u^j} + \frac{\partial g_{ji}}{\partial u^k}\right], \tag{5.6-4}$$

and the *Christoffel symbol of the second kind* as

$$\left\{\begin{matrix} m \\ jk \end{matrix}\right\} = \sum_{(i)}g^{im}[jk, i]. \tag{5.6-5}$$

These symbols are functions of u^1 and u^2 on the surface; they are symmetric in the indices j, k. The equations of a geodesic are then

$$\ddot{u}^m + \sum_{(j,k)} \left\{ \begin{matrix} m \\ jk \end{matrix} \right\} \dot{u}^j \dot{u}^k = 0 \qquad (m = 1, 2) \qquad (5.6\text{-}6)$$

regarded as differential equations for the functions $u^1(s)$ and $u^2(s)$, which describe a geodesic in terms of the arclength s. The vector

$$\vec{w} = \begin{pmatrix} \dot{u}^1 \\ \dot{u}^2 \end{pmatrix} = \vec{w}(s)$$

is a unit vector tangent to the geodesic, that is, "lying along" the geodesic.

Now suppose that $\vec{V} = \vec{V}(s)$ is a two-component vector defined on the geodesic, which satisfies the equation

$$\dot{V}^m + \sum_{(j,k)} \left\{ \begin{matrix} m \\ jk \end{matrix} \right\} \dot{u}^j V^k = 0. \qquad (5.6\text{-}7)$$

Simple calculations show that the quantities $(\vec{V}, \vec{V})$ and $(\vec{V}, \vec{w})$ are constant along the geodesic, where generally $(\vec{V}, \vec{W})$ denotes the inner product given by (5.5-4), namely, $\sum_{(j,k)} V^j g_{jk} W^k$. The constancy of $(\vec{V}, \vec{V})$ shows that if $\vec{V}$ is a unit vector at any point of the geodesic, it is a unit vector at every point, and the constancy of $(\vec{V}, \vec{w})$ shows that the angle between $\vec{V}$ and the tangent vector $\vec{w}$, that is, between $\vec{V}$ and the geodesic, is a constant. This is the analogue of the relation indicated by Fig. 5.6. We therefore say that $\vec{V} = \vec{V}(s)$ is *carried by parallel transport along the geodesic*.

Lastly, suppose that $u^1(s), u^2(s)$ describe a smooth curve, not necessarily a geodesic, where s is arclength, and that $\vec{V}(s)$ is a vector on the curve that satisfies (5.6-7). We then say that $\vec{V}(s)$ is *carried along the curve by parallel transport*.

The general problem of parallel transport along a given curve is an initial-value problem consisting of the differential equation (5.6-7) and an initial condition $\vec{V}(s_0) = \vec{V}_0$ (given). If $\left\{ \begin{matrix} m \\ jk \end{matrix} \right\}$ and u^j are smooth enough functions of s on the curve, this initial-value problem then determines $\vec{V}(s)$ on the curve, and we say that the initial vector $\vec{V}_0$ has been carried along the curve by parallel transport.

We now return briefly to the problems of geodesics. Since the differential equations (5.6-6) are of the second order, there are two types of problems of geodesics: initial-value problems and two-point boundary-value problems. In the initial-value problems, the values of $u^j(0)$ and their first derivatives $\dot{u}^j(0)$ are given, and the problem is then to solve the equations for $u^j(s)$ for $s > 0$. The initial values must satisfy the relation

$$\dot{u}^1(0)^2 + \dot{u}^2(0)^2 = 1,$$

assuming that the parameter s is arclength.

For the two-point problems, the problem is to find a geodesic going from a point P, which has known values of the coordinates to another point Q, which also has known values of the coordinates. In this case, it is not convenient to take the arclength s as the independent variable, because the range of s is not known in advance unless we know the distance between the two points. Instead, we take as independent variable $t = \text{const} \cdot s$. The same differential equations (5.6-6) hold, and the boundary values are $u^j(t_1)$ $(j = 1, 2)$ and $u^j(t_2)$, where t_1 and t_2 are arbitrary; they could be taken as $t_1 = 0$ and $t_2 = 1$.

Since the differential equations are nonlinear, we cannot say much in advance about the range of the solutions. In the general case of a surface with a metric, the solution of the initial-value problem may not exist for all $s > 0$, and the solution of the two-point problem may not be unique, unless the points P and Q are close enough together. However, we state without proof that in the hyperbolic plane, in terms of the coordinate systems used in the models in Chapter 7, the solution of the initial-value problem exists for all s, positive and negative, and the solution of the two-point problem exists and is unique, for all choices of P and Q.

Exercises

1. Show that if $\overrightarrow{V}$ and $\overrightarrow{W}$ are vectors that are carried along the same curve by parallel transport, then the inner product $(\overrightarrow{V}, \overrightarrow{W})$ is constant along the curve.

2. Show that if $u(s), v(s)$ denote a closed curve, for $a \leq s \leq b$, that is, $u(b) = u(a)$ and $v(b) = v(a)$, then the effect of parallel transport round the curve is to rotate $\overrightarrow{V}$ through some angle. Find an expression for $\overrightarrow{V}(b) - \overrightarrow{V}(a)$ as an integral.

3. What sort of smoothness of the curve and of the metric on the surface are necessary for the initial-value problem to be well defined?

4. Discuss the case in which the curve has one or more simple corners, and show that if a vector is carried by parallel transport round a triangle or polygon, the angle through which it is turned is equal to the angular defect or angular excess of the triangle or polygon.

5. Find the equations of a geodesic and the equations of parallel transport for a curve in the Euclidean plane, with the coordinates u, v taken as the polar coordinates r, θ. Do the same for the surface of a sphere, with u, v taken as the coordinates φ and θ of the spherical coordinate system r, φ and θ.

6. Show that if u, v are Cartesian coordinates in the Euclidean plane, then the three-index symbols are zero; hence if $\vec{V}(s)$ is carried by parallel transport on any curve, the (Cartesian) components of $\vec{V}(s)$ are constants.

7. Compute the three-index symbols for the hyperbolic plane, when the coordinates u, v are the coordinates ξ, η introduced in Section 3.9.

5.7 Approximate Laws for Very Small Right Triangles

In Section 4.9 we considered a plane tangent to a surface (a horosphere), both embedded in $\mathbb{H}^3$; from the Euclidean geometry in the horosphere we deduced the locally approximately Euclidean geometry in the hyperbolic tangent plane. Here we consider briefly the opposite situation. We consider a plane tangent to a smooth surface, both embedded in Euclidean space $\mathbb{E}^3$; from the Euclidean geometry in the plane, we deduce the locally approximately Euclidean geometry in the surface. The conclusions hold also for an abstract surface as in Section 5.3, provided that the $g_{jk}(u, v)$ are sufficiently smooth functions of u and v; in any case, we assume that they are at least continuously differentiable functions, so that geodesics are well defined; see Equations (5.6-4)–tan(5.6-6).

We omit the proofs, which are similar to those in Section 4.9.

Let ABC be a small triangle in the surface, with geodesics as sides and with angles as determined in Section 5.5, the angle at vertex C being a right angle. Let a, b, c be the lengths of the geodesic segments BC, AC, AB, as determined in Section 5.2. Then, it can be proved,

$$a = (c \sin \alpha)(1 + \varepsilon), \qquad b = (c \cos \alpha)(1 + \varepsilon')$$

$$a^2 + b^2 = c^2(1 + \varepsilon''),$$

where $\varepsilon, \varepsilon', \varepsilon''$ are quantities that $\to 0$ as $c \to 0$.

Exercise

1. Express ε'' in terms of ε and ε'.

5.8 Area

Areas of polygons in the hyperbolic plane were discussed in Section 3.12. Here, we start anew and derive independently a formula for the area of a region on a surface embedded in 3-space. The formula is expressed in terms of the metric tensor, and we shall see (in the next chapter) that when the metric tensor is as derived in Section 5.3 for the hyperbolic plane, we have agreement with Section 3.12; namely, if the unit of length is such that $K = 1$, the area of a polygon is equal to its angular defect.

We consider first the case where u, v are *orthogonal* curvilinear coordinates in the surface S, so that the metric tensor G is

$$G = \begin{pmatrix} g_{11} & 0 \\ 0 & g_{22} \end{pmatrix} = \begin{pmatrix} g_{11}(u, v) & 0 \\ 0 & g_{22}(u, v) \end{pmatrix}.$$

Let Δu and Δv be small increments of u and v. The corresponding small segments in the surface are perpendicular, and their lengths are, according to Section 5.2,

$$(\Delta s)_1 \approx \sqrt{g_{11}}\, \Delta u \text{ and } (\Delta s)_2 \approx \sqrt{g_{22}}\, \Delta v.$$

The area of the resulting rectangle is

$$\Delta A \approx (\Delta s)_1 (\Delta s)_2 \approx \sqrt{g_{11} g_{22}}\, \Delta u \Delta v = \sqrt{\det G}\, \Delta u \Delta v.$$

Hence if R is a region on S corresponding to a region Ω of the u, v parameter plane, we *define* the area of R by the equation

$$\text{Area}(R) = \int\int_\Omega \sqrt{\det G}\, du dv. \tag{5.8-1}$$

This result holds also for general nonorthogonal coordinates, because the expression on the right of the above equation is invariant under a change of coordinates, say from u, v to u', v'. Let K be the Jacobian matrix defined in Section 5.2, namely,

$$K = \begin{pmatrix} \dfrac{\partial u'}{\partial u} & \dfrac{\partial u'}{\partial v} \\ \dfrac{\partial v'}{\partial u} & \dfrac{\partial v'}{\partial v} \end{pmatrix} = J^{-1}.$$

It follows from (5.2-10) that the metric tensors G and G' for the two coordinate systems are related by

$$G = K^T G' K, \tag{5.8-2}$$

hence,

$$\det G = \det G' (\det K)^2,$$

so that

$$\text{Area}(R) = \int\int_\Omega \sqrt{\det G'}\, |\det K| du dv.$$

According to the chain rule for changing variables in a double integral, this is precisely

$$\text{Area}(R) = \int\int_{\Omega'} \sqrt{\det G'}\, du' dv', \tag{5.8-3}$$

where Ω' is the region of variation of $u'v'$ that corresponds to the region Ω of variation of u, v. Since (5.8-3) is of the same form as (5.8-1), we see that the above definition of area is independent of the particular choice of curvilinear coordinates on $\mathcal{S}$.

The application of this formula to triangles in the hyperbolic plane will be carried out in the next chapter.

Exercise

1. Find the area of a circular region on the unit sphere. *Hints:* Assume that the center of the circle is at the north pole, and explain why that is justified. Express the result in terms of the radius of the circle, defined as the great-circle distance from the center to points on the circle.

5.9 The Gaussian Total Curvature of a Surface

We recall that the angular *excess* of a triangle on the surface of a sphere of radius R is equal to A/R^2 where A is the area of the triangle. According to Section 3.12, the angular *defect* of a triangle in the hyperbolic plane is equal to AK^2, where A is the area of the triangle and K is a constant with dimension $(\text{length})^{-1}$. We combine the two statements by calling the angular defect a negative value of the excess, and then saying that in both cases the angular excess is equal to $\mathcal{K}A$, where $\mathcal{K}$ is equal to R^{-2} in the spherical case and is equal to $-K^2$ in the hyperbolic case; $\mathcal{K}$ is called the *Gaussian* or *total curvature.*

We now consider an abstract surface, in the sense of Section 5.3, and we define a quantity $\mathcal{K}$ in terms of the metric, which reduces to R^{-2} for the sphere and to $-K^2$ for the hyperbolic plane; in general it can vary from point to point in the surface. When it is constant, the surface represents a sphere, a Euclidean plane, or a hyperbolic plane, according as the value is positive, zero, or negative. A surface of constant curvature is, at least locally, one of those three.

When $\mathcal{K}$ is not constant, it has the interpretation of the ratio of angular excess to area for very small triangles, in the limit as the size tends to zero. The same is true for very small general polygons.

We derive the formula for $\mathcal{K}$, following Gauss, first in a polar coordinate system, and then we give the formula in a general coordinate system. The metric is given by

$$ds^2 = dr^2 + g_{22}d\theta^2, \qquad G = \begin{pmatrix} 1 & 0 \\ 0 & g_{22} \end{pmatrix}. \qquad (5.9\text{-}1)$$

Such a coordinate system can always be chosen, at least locally: We choose an arbitrary origin O; for any point P, r is the arclength from O to P along the geodesic through the points O and P, and θ is the angle at the origin

O between the tangent vector of that geodesic and some arbitrary fixed vector at O, counted as positive in the counter-clockwise direction (right-hand rule). It is easy to see (Exercise 1 below) that r, θ is an orthogonal system, so that G is diagonal, as above. The entry g_{22} depends generally on both r and θ, but is always positive.

Let $r(s), \theta(s)$ be a geodesic (not necessarily through the origin), as in Fig. 5.9a, with s the arclength parameter. Consider the vectors

$$\vec{v} = \begin{pmatrix} \dot{r} \\ \dot{\theta} \end{pmatrix}, \quad \vec{w}_r = \begin{pmatrix} \dot{r} \\ 0 \end{pmatrix}, \quad \vec{w}_\theta = \begin{pmatrix} 0 \\ \dot{\theta} \end{pmatrix}$$

at some point P of the geodesic; $\vec{v}$ is a unit vector along the geodesic; $\vec{w}_r$ and $\vec{w}_\theta$ are vectors (not generally unit vectors) at P in the directions of increasing r for fixed θ and of increasing θ at fixed r. They are orthogonal.

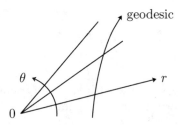

Fig. 5.9a.

Let ω be the angle between $\vec{v}$ and $\vec{w}_\theta$. According to Section 5.5,

$$\cos \omega = \frac{\vec{v}^T G \vec{w}_\theta}{(\vec{w}_\theta^T G \vec{w}_\theta)^{1/2}} = \sqrt{g_{22}} \dot{\theta}. \tag{5.9-2}$$

We assume the sign of ω so chosen that it represents the angle from $\vec{v}$ to $\vec{w}_\theta$, taken as positive in the counter-clockwise direction. Then, from the right-handedness of the r, θ system and its orthogonality, we see that similarly,

$$\sin \omega = \sqrt{g_{11}} \, \dot{r} = \dot{r}. \tag{5.9-3}$$

We differentiate this equation with respect to arclength s along the geodesic, to give $(\cos \omega) \dot{\omega}$ and then we divide by (5.9-2):

$$\dot{\omega} = \frac{\ddot{r}}{\sqrt{g_{22}} \, \dot{\theta}}.$$

It is easily seen that in the coordinate system r, θ the first of the equations (5.6-6) for a geodesic becomes

$$\ddot{r} = \frac{1}{2} \frac{\partial g_{22}}{\partial r} \dot{\theta}^2,$$

hence

$$\dot{\omega} = \frac{1}{2\sqrt{g_{22}}} \frac{\partial g_{22}}{\partial r} \dot{\theta} = \frac{\partial \sqrt{g_{22}}}{\partial r} \dot{\theta}. \tag{5.9-4}$$

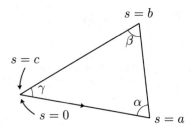

Fig. 5.9b.

We now let $r(s), \theta(s)$ be the path around a triangle, consisting of three pieces of geodesics, corresponding to the intervals $(0, a), (a, b)$, and (b, c) of the arclength s, as indicated in Fig. 5.9b. We now have to distinguish between $\omega(s^-)$ and $\omega(s^+)$ at the vertices, where ω decreases discontinuously by an amount equal to the exterior angle of the triangle. That is, if α, β, γ are the interior angles, as in Fig. 5.9b,

$$\omega(a^-) - \omega(a^+) = \pi - \alpha,$$
$$\omega(b^-) - \omega(b^+) = \pi - \beta,$$
$$\omega(c^-) - \omega(c^+) = \pi - \gamma.$$

Now, Equations (5.9-2) and (5.9-3) determine ω only modulo 2π. Although $\omega(c^+)$ and $\omega(0^+)$ are the same (mod 2π), ω has decreased by 2π in going around the triangle, hence

$$\omega(c^-) - \omega(0^+) = -\pi - \gamma.$$

Integrating (5.9-4) from $s = 0$ to $s = c$ gives $[\omega(a^-) - \omega(0^+)] + [\omega(b^-) - \omega(a^+)] + [\omega(c^-) - \omega(b^+)]$. Therefore,

$$\oint_{\text{triangle}} \frac{\partial \sqrt{g_{22}}}{\partial r} d\theta = \pi - (\alpha + \beta + \gamma) = \text{angular defect} = -\text{angular excess}.$$

By Green's theorem in the plane, we therefore have

$$\text{Angular excess} = -\int\int_{\text{triangle}} \frac{\partial^2 \sqrt{g_{22}}}{\partial r^2} \, dr d\theta.$$

We define

$$K = -\frac{1}{\sqrt{g_{22}}} \frac{\partial^2 \sqrt{g_{22}}}{\partial r^2}, \tag{5.9-5}$$

and we note that for the metric (5.9-1) $g_{22} = \det G$. Finally, then,

$$\text{Angular excess} = \int \int_{\text{triangle}} K\sqrt{\det G} \, dr d\theta. \tag{5.9-6}$$

This is the final formula, and it also holds for a general polygon.

According to the preceding section, $\sqrt{\det G} \, dr d\theta$ is the element of area. Therefore, in the special case in which K is a constant, we have that the angular excess of a polygon is equal to K times its area, as stated at the beginning of this section.

Lastly, if u, v are general coordinates (sometimes called u^1 and u^2), the formula for K is

$$\begin{aligned}
K = \frac{1}{\det G} &\left(\frac{\partial^2 g_{12}}{\partial u \partial v} - \frac{1}{2} \frac{\partial^2 g_{11}}{\partial v^2} - \frac{1}{2} \frac{\partial^2 g_{22}}{\partial u^2} \right. \\
&\left. + \sum_{(p,q)} g^{pq} \Big[[12,p][12,q] - [11,p][22,q] \Big] \right),
\end{aligned} \tag{5.9-7}$$

where $[ij,k]$ are the Christoffel symbols defined in (5.6-4), and the g^{pq} are the entries of the reciprocal of the matrix G, as in (5.6-3).

Exercises

1. Verify that the coordinates r, θ defined in the first paragraph are orthogonal, as claimed in (5.9-1). *Hint:* From the definition of r, the points
    $$(0,0), (r, \theta), (r, \theta + \Delta\theta)$$
    are the vertices of a thin isosceles triangle.
2. Evaluate K from (5.9-5) for the sphere.
3. Evaluate K for the hyperbolic plane. *Hint:* Use the metric in the form given by (7.8-2) in Chapter 7, with r replaced by Kr.
4. Show that (5.9-7) reduces to (5.9-5) for the metric (5.9-1).
5. Verify that the formula (5.9-7) for the curvature is invariant under a transformation of coordinates $u, v \rightarrow u', v'$. *Hint:* First read next section.
6. Evaluate the curvature K for various points of the surface (paraboloid) given by $(x/a)^2 + (y/b)^2 + (z/c)^2 = 1$, and give a geometric interpretation.
7. For a surface embedded in Euclidean 3-space, K is the *total curvature* defined as follows: Any geodesic passing through a point P of the surface, as a curve in space, has a certain radius of curvature at P, ρ. The center of curvature is a point on the normal to the surface at P. One side of the surface is chosen arbitrarily and ρ is considered as positive

or negative according as the center lies on that side or on the other side. As the geodesic varies among all geodesics through P, ρ has a maximum value ρ_1 and a minimum value ρ_2. Then the total curvature at P is defined as $1/\rho_1\rho_2$. Show that

$$\mathcal{K} = \frac{1}{\rho_1\rho_2}.$$

8. For a surface embedded in 3-space, if $\mathcal{K}$ is identically zero, what are the possible shapes of the surface?

9. Show that if $\mathcal{K} > 0$, then in some neighborhood of P the surface lies entirely on one side of the tangent plane, while if $\mathcal{K} < 0$, then in any neighborhood of P, the surface lies partly on one side of the tangent plane and partly on the other side.

5.10 Differentiable Surfaces

The purpose of this section is to prepare a context for formulating the geometric notions of the earlier sections of the chapter so that they are independent of the choice of coordinate systems. We also show how to use coordinates to transfer calculations to the parameter spaces. The most fundamental issue is how to tell whether a given function is differentiable, so we work toward a definition of differentiability.

To begin with, we return to parametrized surfaces. If $\mathcal{S}$ is a surface embedded in $\mathbb{R}^3$ and parametrized as in Section 5.1, let φ be the function from Ω to $\mathbb{R}^3$ whose components are X, Y, and Z. Then the inverse function φ^{-1} is continuous and maps $\mathcal{S}$ one-to-one onto Ω. Hence a function $f : \mathcal{S} \to \mathbb{R}$ is continuous if and only if $f \circ \varphi$ is continuous on Ω.

What about differentiability of f? The function $f \circ \varphi$ can generally be no smoother than φ, so we investigate the general case in which we know that φ is of class C^k, where $k \geq 1$. (That means that all partial derivatives of order at most k exist and are continuous.) We recall the standard notion of smoothness of a function on a surface in $\mathbb{R}^3$: If $f : \mathcal{S} \to \mathbb{R}$ and $P \in \mathcal{S}$, then f is of class C^n near P on $\mathcal{S}$ if and only if f has an extension to a neighborhood of P that is of class C^n on the neighborhood. If $n \leq k$, it is clear that this condition implies that $f \circ \varphi$ is of class C^n in a neighborhood of $\varphi^{-1}(P)$. To prove the converse we need to use the Inverse Function Theorem.

Suppose that $P = \varphi(a, b)$ and let γ_1 and γ_2 be the straight line parametric curves in Ω through (a, b) in the first and second coordinate directions. Then the tangent vectors to the curves $\varphi \circ \gamma_1$ and $\varphi \circ \gamma_2$ at P are tangent to the surface $\mathcal{S}$ and their cross product, V, is normal to $\mathcal{S}$. Let W be the tangent plane to $\mathcal{S}$ at P, which is the same as the plane through P and perpendicular to V. Let ψ be the function on $\mathbb{R}^3$ that projects points onto W parallel to V. The plane W is the image of a two dimensional sub-space, W_0, of $\mathbb{R}^3$, under a translation τ that takes 0 to P. Let T be gotten by composing some vector space isomorphism of W_0 onto $\mathbb{R}^2$ with $\tau^{-1}|W$. Then T is one-to-one and onto, and $T \circ \psi$ is of degree one in terms of the standard coordinates on $\mathbb{R}^3$ and $\mathbb{R}^2$. Thus $T \circ \psi$ differs from its derivative only by an additive constant (which is a vector). The derivative $D\varphi(a,b)$ is an isomorphism of $\mathbb{R}^2$ onto W_0 because φ is an embedding, and the derivative of $T \circ \psi$ is one-to-one on W_0. By the chain rule, the derivative of $T \circ \psi \circ \varphi$ at (a,b) is also invertible. By the Inverse Function Theorem, $T \circ \psi \circ \varphi$ maps a neighborhood of (a,b) one-to-one onto a neighborhood of its image. It follows that ψ is one-to-one on the part of S in a small enough neighborhood, N, of P. If $f : S \to \mathbb{R}$, then we can extend f to N by letting the extension be constant on lines parallel to V.

Exercise. Verify that this extension function is of class C^n provided that $n \le k$ and $f \circ \varphi$ is of class C^n. *Hints:* Think about the inverses of $T \circ \psi \circ \varphi$ and T. Think about differentiability of a function on a plane in $\mathbb{R}^3$.

Turning to the abstract setting, let S be a metric space (a set equipped with a distance function, as in the axioms for the hyperbolic plane). The metric is only needed so that it makes sense to assume the existence of a convex region Ω in $\mathbb{R}^2$ and a one-to-one continuous mapping φ from Ω onto S that has a continuous inverse. That means that φ is a *homeomorphism* of Ω onto S, and φ^{-1} is also a homeomorphism of S onto Ω. Our main interest is in the cases in which S is an open disc or an open half-plane, and the Riemannian metric of interest provides a distance function that is definitely different from the intial one. Once again, we know that a function $f : S \to \mathbb{R}$ is continuous if and only if $f \circ \varphi$ is continuous. Our plan is to exploit the existence of the homeomorphism of S with Ω to carry additional structure of Ω over to S. Since S is not given as a subset of $\mathbb{R}^3$ or some other coordinate space, we have no a priori meaning of differentiability of f. There is also no way to say how smooth φ is, giving us considerable freedom at this point. Only when there are two or more coordinate charts does the situation get more constrained, so we start with the following definition.

Definition 1: We say that a function f from S to $\mathbb{R}$ is *of class C^n with respect to φ* if and only if $f \circ \varphi$ is of class C^n, for $1 \le n \le \infty$.

Next we want to transfer the definition of partial derivative from Ω to S. If f is a real valued function on S, write $\hat{f}$ for $f \circ \varphi$, so that $f = \hat{f} \circ \varphi^{-1}$. By our definition of $C^n(S)$, $f \mapsto \hat{f}$ is an isomorphism of $C^n(S)$ onto $C^n(\Omega)$. We can use this connection to define partial derivatives, by deciding to make partial derivatives of f correspond to the "same" derivatives of $\hat{f}$. Let ξ^1 and ξ^2 be the first and second coordinate functions on Ω, and set $x^1 = \xi^1 \circ \varphi^{-1}$, $x^2 = \xi^2 \circ \varphi^{-1}$. Then $\hat{f}(x^1(P), x^2(P)) = f(P)$ for $P \in S$. We define

$$\frac{\partial f}{\partial x^j} = \frac{\partial \hat{f}}{\partial \xi^j} \circ \varphi^{-1},$$

making it true that

$$\left(\frac{\partial f}{\partial x^j}\right)^{\wedge} = \frac{\partial \hat{f}}{\partial \xi^j}.$$

Many other facts are automatic. For example, for smooth enough f, mixed partials are independent of the order in which the derivatives are taken.

For some purposes the notation is more convenient if we reverse our point of view and emphasize the coordinates rather than the parametrization.

Definition 2: Let $\mathcal{M}$ be a metric space. If U is an open set in $\mathcal{M}$ and φ is a homeomorphism of U onto a convex open set in $\mathbb{R}^2$, we say that (U, φ) is a *(two dimensional) coordinate chart* in $\mathcal{M}$. If (U, φ) and (V, ψ) are coordinate charts in $\mathcal{M}$ and $1 \le k \le \infty$ then we say that the two charts are C^k-*compatible* if and only if either $U \cap V = \emptyset$ or $\psi \circ \varphi^{-1}$ is C^k from $\varphi(U \cap V)$ onto $\psi(U \cap V)$ and has a C^k inverse.

Notice that $\varphi(U \cap V)$ and $\psi(U \cap V)$ are open sets in $\mathbb{R}^2$ so that the differentiability condition is meaningful. If (U, φ) and (V, ψ) are C^k-compatible coordinate charts then it makes sense to define a function f on $U \cup V$ to be of class C^n (for $n \le k$) if and only if $f \circ \varphi^{-1}$ and $f \circ \psi^{-1}$ are both of class C^n. It also makes sense to define a function f from $\mathbb{R}$ or $\mathbb{R}^2$ into $U \cup V$ to be of class C^n (for $n \le k$) if and only if $\varphi \circ f$ and $\psi \circ f$ are of class C^n on their domains of definition. These terms are well-defined because on $U \cap V$ the two conditions agree. For the moment, our main interest is in the case when U and V are both equal to $\mathcal{M}$, but it also makes sense to consider arbitrary families of pairwise compatible charts.

Definition 3: Let $\mathcal{M}$ be a metric space. A (2-dimensional) C^k *coordinatization of* $\mathcal{M}$ is a family $\{(U_i, \varphi_i) : i \in I\}$ of C^k-compatible two dimensional coordinate charts in $\mathcal{M}$ such that

$$\mathcal{M} = \bigcup_{i \in I} U_i.$$

A metric space $\mathcal{M}$ is called a C^k *surface* if it has a 2-dimensional C^k coordinatization. If $\mathcal{M}$ is a C^k surface, $n \le k$, and U is an open set in $\mathcal{M}$, we define $C^n(U) = \{f : U \to \mathbb{R} : \text{for every } i \in I \text{ we have } (f|(U \cap U_i)) \circ \varphi_i^{-1}$ of class C^n on $\varphi(U \cap U_i)\}$.

Lemma: If $\mathcal{M}$ is a C^k surface, $n \le k$, and U is an open set in $\mathcal{M}$, then $C^n(U)$ is a vector space over $\mathbb{R}$ and is closed under pointwise multiplication.

Note. The reader should prove this lemma as an exercise in understanding the definitions.

The derivatives of a function at a point depend only on the values of the function near the point. We can say that derivatives are *local* in character. Because of that, many geometric features are also local. We need a tool for reducing to local questions, and find it in the existence of C^∞ functions that vanish outside a prescribed open set but take the value 1 on some neighborhood of a given point of that set.

Lemma: Let U be an open set in $\mathbb{R}^n$ and let $P \in U$. Then there is a $\delta > 0$ and a function $f \in C^\infty(\mathbb{R}^n)$ such that $f(Q) = 0$ for $Q \notin U$ and $f(Q) = 1$ if the distance from Q to P is at most δ.

Proof (Outline): We combine the fact that C^∞ functions can be added and multiplied with some elementary calculus of one variable. It will suffice to prove the result for the case that $n = 1$. The first of several steps in the construction is to consider the function $g : \mathbb{R} \to \mathbb{R}$ defined by

$$g(x) = \begin{cases} \exp(-x^{-2}) \text{for } x > 0 \\ 0 \text{ for } x \le 0 \end{cases}.$$

An induction argument shows that $g \in C^\infty$. If $a > 0$, then $h(x) = g(x)g(a - x)$ defines a C^∞ function that vanishes off $[0, a]$. Next define H by

$$H(x) = \int_0^x h(t)\, dt$$

to get a C^∞ function that is 0 on $(-\infty, 0]$ and some positive constant on $[a, \infty)$. Now suppose that $b < c$. If a and A are suitable positive numbers, then $AH(x - b)H(c - x)$ vanishes off (b, c) and is 1 near $(b + c)/2$.   □

To be explicit about the localization of function behavior, we prove the following lemma.

Lemma: Let $\mathcal{M}$ be a C^k surface, let U be an open set in $\mathcal{M}$ and let $P \in U$. Suppose that $g \in C^k(U)$. Then there exist an $f \in C^k(\mathcal{M})$ and an open set V such that $P \in V \subset U$ and $f|V = g|V$.

Proof: We lose no generality by assuming that U is the open set of a co-ordinate chart. Then we can find a function h that is in C^∞ relative to that coordinate chart, is 0 off some compact set in U that contains P in its interior, and that is identically 1 on some neighborhood V of P. Then hg can be extended to a C^k function on $\mathcal{M}$ by using the value 0 off U. Call that extension f.   □

5.11 Vectors and Tensors

In this section, we present two ways of defining vectors and tensors on a smooth surface. The first one is more common in mathematics. The main idea is that there should exist a well-defined object and we want to be able to convert its components or numerical expression relative to one coordinate system to the corresponding elements relative to another coordinate system. For example, the surface itself is specified as a set and the usual definition of function applies. The second approach is more common in physics. The emphasis is on the coordinates of points, and the behavior of components under changes of coordinates tells what kind of object is under consideration. Thus a function on the surface is called a scalar field: there is a function of the coordinates specified for every coordinate system and the various functions have the same value at coordinates of the same point.

Let us now consider how to define tangent vectors intrinsically. Let M be a C^k surface and let P be a point of M. If (U, φ) is a coordinate chart of M such that $P \in U$, then tangent vectors to M at P should be in one-to-one correspondence with the tangent vectors to $\varphi(U)$ at $\varphi(P)$. Also, a C^1 curve that passes through P should have a tangent vector in the tangent space at P. Essentially all that we have to work with are the smooth functions defined on M or taking values in it. Two uses of vectors in $\mathbb{R}^2$ provide hints of how to proceed. A smooth parametrized curve has a tangent vector at each point that indicates the velocity of motion of a point tracing out the curve. This tangent vector can be calculated as a limit of difference quotients. We can also take the derivative of a smooth function at a point with respect to a vector (the "directional derivative"). Indeed, the chain rule brings these two ideas together if we look at the derivative of a function along a parametrized curve. The interaction between functions, curves and vectors is what we want to exploit in generalizing the notion of tangent vectors.

Let us write S for $\varphi(U)$, let $f \in C^1(S)$, and choose a point $(a, b) \in S$. The derivative $Df(a, b)$ is a linear function from $\mathbb{R}^2$ to $\mathbb{R}$, and its value at (x, y) is defined to be

$$\lim_{t \to 0} \frac{f(a + tx, b + ty) - f(a, b)}{t}.$$

If we abbreviate $(Df(a, b))(x, y)$ by $L(f)$, then L is defined as a function from $C^1(S)$ to $\mathbb{R}$. We can recover (x, y) from L as follows: $x = L(x^1)$ and $y = L(x^2)$, where x^1 and x^2 are the first and second coordinate functions on S. Thus L is a kind of replacement for (x, y), but we need to know what that 'kind' is before we can use it for a definition. The essential clue is that L is computed as a derivative at (a, b), so it is linear and satisfies Leibnitz' rule. For real numbers α and β, and for smooth functions f and g, we have:

$$L(\alpha f + \beta g) = \alpha L(f) + \beta L(g) \qquad (5.10\text{-}1)$$

$$L(fg) = L(f)g(a,b) + f(a,b)L(g) \qquad (5.10\text{-}2)$$

Another property of L that can be proved from these is that if f is a constant function then $L(f) = 0$:

$$L(1) = 0. \qquad (5.10\text{-}3)$$

These three properties all follow just from the fact that the value of $L(f)$ is computed as a limit of a difference quotient. Thus if γ is any smooth curve for which $\gamma(0) = (a,b)$, and we define $L(f) = (f \circ \gamma)'(0)$, then L has those same properties. It can be proved that the only functions from $C^\infty(\mathcal{S})$ to $\mathbb{R}$ that have the basic properties (5.10-1) and (5.10-2) are the ones that arise from differentiating along a curve, but it will be simpler just to use the functions L that are *defined* as derivatives along curves. We leave the proof of the following lemma as an exercise.

Lemma: If γ is a C^1 curve in $\mathcal{S}$ and $\gamma(0) = (a,b)$, write (x,y) for $\gamma'(0)$, the derivative as a limit of difference quotients. Then $(f \circ \gamma)'(0) = (Df(a,b))(x,y)$ for every C^1 function from $\mathcal{S}$ to $\mathbb{R}$.

Definition 4: If $\mathcal{M}$ is a C^k surface and $P \in \mathcal{M}$, we write $T_P(\mathcal{M})$ for the set of functions $L : C^1(\mathcal{M}) \to \mathbb{R}$ that can be written in the form $L(f) = (f \circ \gamma)'(0)$ for some C^1 curve into $\mathcal{M}$ for which $\gamma(0) = P$. The set $T_P(\mathcal{M})$ is called the *tangent space to* $\mathcal{M}$ *at* P. If γ is a curve for which $L(f) = (f \circ \gamma)'(0)$ then we say that L is the *tangent vector to* γ *at* P.

In the physics literature and sometimes in the mathematics literature the term *contravariant vector* is used for what we have called a tangent vector. Let the reader be aware that these vectors are carried forward by mappings between manifolds, in spite of this name.

A coordinate chart (U, φ) such that $P \in U$ gives an easy way to get some tangent vectors. Let x^1 and x^2 be the coordinate functions associated with φ as in Section 5.10, and write $D_j f = \partial f / \partial x^j$ for $j = 1, 2$. Let γ be a C^1 curve in $\mathcal{M}$ such that $\gamma(0) = P$, and suppose that the derivative $(\varphi \circ \gamma)'(0)$ computed as a limit of difference quotients in $\mathbb{R}^2$ is (c_1, c_2). By the definition of partial derivative in (U, φ) and the chain rule, $(f \circ \gamma)'(0) = c_1 D_1 f(P) + c_2 D_2 f(P)$. Let $\varphi(P) = (a,b)$. Since there is a one-to-one correspondence between C^1 curves in U and C^1 curves in $\varphi(U)$, we know that there is always a γ such that $\varphi \circ \gamma(t) = (a + c_1 t, b + c_2 t)$ at least for $|t|$ small enough. Using the definition and working in $\varphi(U)$, we can show that D_1 and D_2 are linearly independent at each P. This reasoning is an example of how the analysis can be transferred to the parameter space. It proves the next lemma.

Lemma: $T_P(\mathcal{M})$ is a subspace of dimension 2 in the vector space of all real valued functions on $C^1(\mathcal{M})$ under pointwise operations.

Once we have a tangent space, we also have a cotangent space. The *cotangent space* is the dual space of $T_P(\mathcal{M})$ as a vector space, denoted by $T_P^*(\mathcal{M})$. Cotangent vectors are also called *covariant vectors*. Notice that if $f \in C^1(\mathcal{M})$ and $P \in \mathcal{M}$, then f determines a cotangent vector at P. Indeed, $L \mapsto L(f)$ is a linear functional on $T_P(\mathcal{M})$ because the vector operations of $T_P(\mathcal{M})$ are defined to be pointwise operations on the elements as functions on $C^1(\mathcal{M})$. In Euclidean space, a directional derivative can be written as the dot product of the gradient with the vector in question. This is reminiscent of that situation, but we use slightly different notation: the cotangent vector determined by f at P is written as df_P. Thus $df_P(L) = L(f)$. In particular, we can define $(dx^1)_P$ and $(dx^2)_P$ in this manner for every $P \in U$, by finding global functions that agree with x^1 or x^2 in some neighborhood of P. These give the basis of $T_P^*(\mathcal{M})$ that is dual to the basis $\{((D_1)_P, (D_2)_P)\}$ of $T_P(\mathcal{M})$.

Next, we want to define tensors. There are at least two other ways to do that but the least complicated way is equivalent to the others so we use it.

To simplify notation in this discussion, we write V for $T_P(\mathcal{M})$, so that the dual space of V, V^*, is the same as $T_P^*(\mathcal{M})$. Then we write $V_{r,s}$ for the cartesian product of r copies of V^* and s copies of V:

$$V_{r,s} = V^* \times \cdots \times V^* \times V \times \cdots \times V.$$

Then a function defined on $V_{r,s}$ can be described as a function of r dual vectors and s vectors. A function of that type is called *multilinear* provided that it is linear as a function of each argument when the other arguments are held fixed. To put it another way, if we assign values to $r + s - 1$ of the variables, we obtain a linear function of the remaining variable. Multilinearity generalizes the bilinearity of inner products.

Definition 5: A multilinear function from $T_P(\mathcal{M})_{r,s}$ to $\mathbb{R}$ is called a *tensor, contravariant of degree r and covariant of degree s, or of degree (r,s)*.

Notice that a tensor of degree $(0,1)$ is simply a cotangent vector. Since there is a natural isomorphism between V and $(V^*)^*$, it is also true that a tensor of degree $(1,0)$ can be identified as a tangent vector. There is a notion of *tensor product of vector spaces* in terms of which all tensors are in tensor product spaces and so made up of sums of products of tangent vectors and cotangent vectors. Our definition avoids that machinery, but notice that if v_1 and v_2 are in V and $f \in V^*$, then we can use them to define a tensor of degree $(2,1)$. If f_1 and f_2 are in V^* and $v \in V$, define

$$A(f_1, f_2, v) = f_1(v_1) f_2(v_2) f(v).$$

It is easy to verify that A is multilinear, and it is consistent with the theory of tensor products to write $v_1 \otimes v_2 \otimes f$ for A.

Finally, we are ready to talk about vector and tensor fields, but again must proceed in stages. Keep in mind that a tangent vector is a function defined on functions.

Definition 6: A *vector field* is a function X on $\mathcal{M}$ such that for each P in $\mathcal{M}$ the value of X at P is in $T_P(\mathcal{M})$. The value of X at P is often denoted by X_P. A vector field X is of class C^n if for every $f \in C^n(\mathcal{M})$ the function $X(f)$ taking P to $X_P(f)$ is in $C^{n-1}(\mathcal{M})$.

It is important to see the form given to a vector field by a coordinate chart. Let (U, φ) be a coordinate chart compatible with the coordinatization of $\mathcal{M}$, and set $\Omega = \varphi(U)$. Take x^1 and x^2 to be the same coordinate functions as before so that φ is just the ordered pair (x^1, x^2). Also recall D_1 and D_2 from before; these are actually vector fields since they operate on functions on U to give functions on U. We define $(D_j)_P f = (D_j f)(P)$. For each P, $(D_1)_P$ and $(D_2)_P$ constitute a basis of $T_P(M)$, so if X is any vector field on $\mathcal{M}$ there must be functions c^1 and c^2 so that $X = c^1 D_1 + c^2 D_2$. Applying the vector field to x^1 and x^2 shows that $c^j = D_j x^j$ for $j = 1, 2$.

We make a similar definition for cotangent vector fields, except that they must be applied to tangent vector fields to test smoothness. If $f \in C^n(\mathcal{M})$, there is an associated cotangent vector field, df, of class C^{n-1} and defined by

$$df_P(X_P) = X_P(f).$$

The cotangent vector fields dx^1 and dx^2 provide a basis of $T_P^*(\mathcal{M})$ for every P. From these basic objects, one can build up the more complicated tensors.

We have discussed differentiation on surfaces, and integration is also meaningful. A cotangent vector field has a well-defined integral along a smooth curve, meaning that the calculation can be done in two different coordinate systems and the same value is obtained. The integral of a function relative to arclength also makes sense.

Definition 7: A tensor field of degree (r, s) is a function T assigning to each point $P \in \mathcal{M}$ a tensor of degree (r, s) over $T_P(\mathcal{M})$. The tensor field T is *of class C^n* if and only if whenever it is applied to vector fields and covector fields of class C^n the resulting function is of class C^{n-1}.

It should be verified that applying this definition to the previously defined cases gives consistent results.

We make particular mention of the notion of metric tensor, by which we mean a field of inner products. Suppose that for each point P we have an inner product G_P on $T_P(\mathcal{M})$. That means that G_P is of degree (0,2), so that it is a function of two tangent vectors. Furthermore, it must be linear in each argument, symmetric, and positive definite. To be useful, it must be true that applying G to a pair of smooth vector fields produces a smooth function.

184 Chapter 5. Geometry of Surfaces

Another important kind of tensor of degree (0,2) is one that is not symmetric but skew symmetric, which are called skew forms of degree 2. For example, if f and g are C^2 functions, then we can define such a tensor field by the formula $X(f)Y(g) - X(g)Y(f)$ for any pair (X, Y) of vector fields. Skew forms of degree 2 can be integrated over surfaces, and their analogues of higher degree can be integrated over higher dimensional manifolds. The reader is encouraged to study this topic further in one of the many books treating analysis on manifolds.

After all this abstract mathematics, it may help the reader to remember that one still computes with coordinates. Since there may be a need for more than one coordinate system, it is important to know how the components of vectors, covectors and tensors in general will change from one to another. Suppose that (U, φ) and (V, ψ) are compatible charts, with coordinate functions x^j and y^j respectively. We suppose that the coefficients relative to the x^j are known and compute from them the coefficients relative to the y^j. Recall that $\partial y^j/\partial x^k$ is defined to be a derivative of $y^j \circ \varphi^{-1}$. Hence the matrix of the derivative of $\psi \circ \varphi^{-1}$ has its (j, k) entry equal to $\partial y^j/\partial x^k$. If the tangent vector to γ has components (a^1, a^2) relative to φ and (b^1, b^2) relative to ψ, the chain rule gives

$$ b^j = \frac{\partial y^j}{\partial x^1} a^1 + \frac{\partial y^j}{\partial x^2} a^2 $$

for $j = 1, 2$. The chain rule also applies to show that a covector's coordinates change using the matrix whose (j, k) entry is $\partial x^j/\partial y^k$.

In the physics literature, an assignment of coordinates to all coordinate systems so that the first rule holds is called a contravariant vector. Thus our definition is consistent with that one. A function on $\mathcal{M}$ is just a way to assign numbers to coordinates of points so that the value depends only on the point, not its coordinates. In the physics literature a function is called a *scalar field*.

In our notation above, we can use the coordinate system given by φ to define a vector field: $x^1 D_1 + x^2 D_2$. Changing to the coordinates given by ψ, the coefficients are not y^1 and y^2. (See Exercise 1 below.) One says that (x^1, x^2) "are not the components of a vector."

Now suppose that G and H are the matrix functions giving the coordinates of a metric tensor relative to (x^1, x^2) and (y^1, y^2). Then we get

$$ h_{ij} = \sum_{k,\ell=1}^{2} \frac{\partial y^k}{\partial x^i} \frac{\partial y^\ell}{\partial x^j} g_{k\ell}. $$

This is the way coordinates of tensors of degree (0,2) for two different coordinate charts are related.

Remark: The following discussion and exercises are formulated to cover essentially the same material as the first part of this section from the perspective of the physics literature.

A familiar example of a vector is the tangent vector to a curve. If $u^1(t)$ and $u^2(t)$ describe a curve in a surface in terms of a parameter t, then the quantities $(d/dt)\,u^1(t), (d/dt)\,u^2(t)$ are the components of a *tangent vector* to the curve *with respect to the coordinate system* u^1, u^2. If we make a change of coordinates from u^1, u^2 to u'^1, u'^2, and if functions $u'^1(t), u'^2(t)$ represent the same curve in terms of the same parameter t, then the quantities $(d/dt)\,u'^1(t)$, $(d/dt)\,u'^2(t)$ are said to be the components of the *same* vector with respect to the new coordinates. The relation between the two sets of components was given in Section 5.2; in the present notation, it is

$$\frac{d}{dt}\,u'^j = \sum_{k=1}^{2} \frac{\partial u'^j}{\partial u^k} \frac{d}{dt}\,u^k, \tag{5.11-1}$$

where u'^1 and u'^2 are understood to be the functions $u'^1(u^1, u^2)$ and $u'^2(u^1, u^2)$ according to the convention mentioned in that section. A vector whose components transform in this way is called *contravariant*.

Definition 8: A *contravariant vector* at a given point P of a surface is the association with each coordinate system of a pair w^1, w^2 of numbers so that the transformation law

$$w'^j = \sum_{k=1}^{2} \frac{\partial u'^j}{\partial u^k}\bigg|_P w^k \tag{5.11-2}$$

holds.

Note: If one wishes to have something concrete in mind, one may think of t as time and $u^1(t), u^2(t)$ as the position of a particle moving in the surface. Then the tangent vector is the particle's *velocity*. Equation (5.11-1) tells how the velocity components transform under a rotation of coordinates or a transformation from a system based on inches to one based on centimeters, or a transformation from Cartesian to polar coordinates.

Definition 9: A *contravariant vector field* on a surface is the specification of a contravariant vector at each point P of the surface. (Then the components are functions of u^1 and u^2.)

Clearly these definitions can be extended to any number n of dimensions, where there are n coordinates and n components of a vector, and where the summation in (5.11-2) goes from $k = 1$ to $k = n$.

An example of a contravariant vector field is the velocity field of a fluid flowing in space.

Definition 10: A *scalar field* on a surface is the association with each coordinate system of a function in such a way that the functions all have the same value at a given point P in the surface, so that the transformation law is

$$f'(u'^1, u'^2) = f(u^1, u^2). \tag{5.11-3}$$

The interpretation of this equation is that it becomes an identity if the left member is written as $f'\left(u'^1(u^1, u^2), u'^2(u^1, u^2)\right)$ or if the right member is written similarly in terms of the primed variables.

An example of a *covariant* vector field is the gradient of a scalar field $w_1 = (\partial f/\partial u^1)$, $w_2 = (\partial f/\partial u^2)$.

Definition 11: A *covariant vector field* on a surface is the association with each coordinate system of two functions $w_1(u^1, u^2)$ and $w_2(u^1, u^2)$ so that the transformation law

$$w'_j = \sum_{k=1}^{2} \frac{\partial u^k}{\partial u'^j} w_k \tag{5.11-4}$$

holds.

Contrast this with (5.11-2) by noting where the prime appears on the right hand side and how the summation index k appears.

Note: Contrary to appearances, the coordinates u^1, u^2 do *not* constitute the components of a contravariant vector field (see Exercise 1 below).

Definition 12: A *covariant tensor field of degree two* is the association of four functions $w_{11}, w_{12}, w_{21}, w_{22}$ with each coordinate system so that the following transformation law holds

$$w'_{ij} = \sum_{k,\ell=1}^{2} \frac{\partial u^k}{\partial u'^i} \frac{\partial u^\ell}{\partial u'^j} w_{k\ell}. \tag{5.11-5}$$

It follows from (5.2-8) that the quantities g_{ij} transform as a covariant tensor of degree two. That justifies calling the matrix

$$G = \begin{pmatrix} g_{11} & g_{12} \\ g_{21} & g_{22} \end{pmatrix}$$

the *metric tensor*.

It is evident how to define covariant and contravariant tensors of higher degree, and also mixed tensors, such as $T_k^{ij\ell}$, which is contravariant of degree 3 and covariant of degree 1.

We now state some relations whose proofs make easy exercises.

(A) If v^j is a contravariant vector and w_j is a covariant one, then the quantity $\sum_{j=1}^{2} v^j w_j$ is a scalar.

(B) If v^j and w^j are contravariant vectors and T_{jk} is a covariant tensor of degree 2, then the quantity $\sum_{(j,k)} v^j T_{jk} w^k$ is a scalar.

(C) If v^j is a contravariant vector and g_{jk} is a covariant tensor of degree 2 — for example, the metric tensor — then the quantities $\sum_{j-1}^2 v^j g_{jk}$ $(k = 1, 2)$ are the components of a covariant vector.

(D) If h^{ij} is a contravariant tensor of degree 2 and g_{jk} is a covariant tensor of degree 2, then the quantities $T_k^i = \sum_{j=1}^2 h^{ij} g_{jk}$ are the components of a mixed tensor, covariant of degree 1 and contravariant of degree 1.

(E) If v^j and w^k are contravariant vectors, then the four quantities $v^j w^k$ are the components of a contravariant tensor of degree 2.

Exercises

1. Explain why the coordinates u^1, u^2 are not the components of a vector.
2. Consider the n-dimensional case with coordinates $x^1, x^2, \ldots, x^n$. Show that the n^4 quantities

$$R_{ijk\ell} = \frac{1}{2}\left(\frac{\partial^2 g_{i\ell}}{\partial x^j \partial x^k} + \frac{\partial^2 g_{jk}}{\partial x^i \partial x^\ell} - \frac{\partial^2 g_{ik}}{\partial x^j \partial x^\ell} - \frac{\partial^2 g_{j\ell}}{\partial x^i \partial x^k}\right) \tag{5.11-6}$$
$$+ g^{pq}([jk,p][i\ell,q] - [ik,p][j\ell,q])$$

   are the components of a fourth-degree covariant tensor. It is called the *Riemann tensor* or the *Riemann curvature tensor*. See (5.6-4).
3. Show that the components of the Riemann tensor satisfy the identities:

$$R_{ijk\ell} = -R_{jik\ell} = -R_{ij\ell k} = R_{k\ell ij}. \tag{5.11-7}$$

4. Show that the quantities $R_{jk} = \sum_{(i,\ell)} g^{i\ell} R_{ijk\ell}$ are the components of a covariant tensor of degree 2. It is called the *Ricci tensor*, and the quantity $R = \sum_{(j,k)} g^{jk} R_{jk}$ is a scalar called the *Riemann curvature scalar*.
5. Show from (5.11-7) that in the two-dimensional case $(n = 2)$, all the nonzero components of the Riemann tensor $R_{ijk\ell}$ are equal, except for signs.
6. Show that for $n = 2$ R is equal to the Gaussian curvature $\mathcal{K}$ given by (5.9-7). That shows that the value of $\mathcal{K}$ is independent of the choice of the coordinate system.
7. Show that if to each coordinate system there is associated a pair v^1, v^2 such that for all covariant vectors w_j, the quantity $\sum_{j=1}^2 v^j w_j$ is a scalar, then the quantities v^j are the components of a contravariant vector. Generalize.
8. Show that if g^{jk} are the entries of the reciprocal of the matrix G, as above, then they are the components of a contravariant tensor of degree 2.
9. For a given contravariant vector v^j, the covariant vector $\sum_{j=1}^2 v^j g_{jk}$, where (g_{jk}) is the metric tensor, is often written as v_k and is called the

result of *lowering the index*. Show how to raise the index, and show that the result of raising and then lowering gives back the original vector. In physical applications, the contravariant vector v^j and the covariant vector v_j are usually regarded as just two different ways of representing the same physical quantity.

5.12 Invariance of the Line Element Under Isometries

If we change from u, v to new curvilinear coordinates u', v' by equations

$$u = u(u', v'), \qquad v = v(u', v'), \tag{5.12-1}$$

then the line element ds, given by

$$ds^2 = g_{11}(u, v)du^2 + 2g_{12}(u, v)dudv + g_{22}dv^2 \tag{5.12-2}$$

is changed for two reasons: (1) we have to replace the variables u, v in the functions $g_{jk}(u, v)$ by $u(u', v')$ and $v(u', v')$; (2) we have to replace du by $(\partial u/\partial u')\, du' + (\partial u/\partial v')\, dv'$, and so on. The result is of the form

$$ds^2 = h_{11}(u', v')du'^2 + 2h_{12}(u', v')du'dv' + h_{22}(u', v')dv'^2. \tag{5.12-3}$$

Recall that in Section 3.6 the isometries of the hyperbolic plane were introduced in terms of two sets of polar coordinates, first r, θ with respect to an origin Z and an axis ℓ, and then r', θ' with respect to an origin Z' and axis ℓ', as in Fig. 5.12a. We consider a mapping of the hyperbolic plane onto itself in which a point P with coordinates r, θ is mapped onto a point Q with the same values of the new coordinates r', θ' (i.e., $r' = r$, $\theta' = \theta$). We proved that that mapping is an isometry; it maps any figure onto a congruent figure, so that the entire geometry is invariant under the mapping.

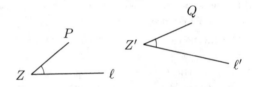

Fig. 5.12a.

If the mapping of a surface onto itself obtained in a similar way from (5.12-1) is an isometry, then the line element must be invariant. That is, the h_{jk} in (5.12-2) must be the same functions of u' and v' as the g_{jk} in (5.12-2) are of u and v. Conversely, if the line element is invariant, the mapping is an isometry, because the entire geometry can be obtained from the line element.

These ideas are used in Section 7.4 to obtain expressions for the isometries of the hyperbolic plane in terms of the Poincaré half-plane coordinates.

Chapter 6

Quantitative Considerations

By use of material in the two preceding chapters, quantitative formulas are derived for many of the objects and relations described qualitatively in Chapter 3. First, a formula is given for the angle of parallelism, which tells in what direction a line m has to be started from a point A to be asymptotic to another line ℓ; it is expressed as a function of the distance y from A to ℓ, obtained by dropping a perpendicular from A to ℓ. From that, an equation in polar coordinates is derived for a horocycle, which is in a sense the limit of a circle as its radius goes to infinity. Next, differential equations and formulas (the latter obtained by solving the differential equations) are obtained for the functions $g(r)$ and $f(r)$ in terms of which the length of a circular arc of radius r and the area of a circular sector of radius r were expressed in Chapter 3. Then, formulas are derived for the legs a, b of a right triangle (not assumed small) of hypotenuse c and one acute angle. The formulas contain the hyperbolic functions sinh and tanh of a, b, and c. A generalization of the law of cosines is found, which gives the length of the third side of a general triangle in terms of two sides a, b and included angle. A formula for a line in polar coordinates is given, and the equation for an equidistant. An equidistant is a curve at a fixed distance from a given line (obtained by dropping perpendiculars); an equidistant is not itself a (straight) line. Ideal points at infinity are defined, and the chapter closes with formulas for certain isometries (especially translations) in polar coordinates.

6.1 The Angle of Parallelism and Horocycles

Let ξ, η be the coordinates described in Section 3.9, and let ℓ be the line through the origin O and perpendicular to the ξ axis, as shown in Fig. 6.1. The equations of ℓ are, according to (5.4-9), with $\xi_0 = 0$ and $\eta_0 = 0$,

$$\xi(s) = \ln \cosh s$$
$$\eta(s) = \tanh s,$$

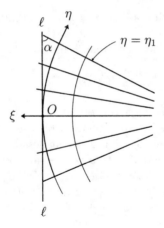

Fig. 6.1.

where s is the distance along ℓ from O. If α is the angle of intersection of ℓ with a line $\eta = \eta_1$ of the asymptotic pencil, as shown, then, according to Section 3.1, $\alpha = \Pi(s)$, where s is the distance from O to the intersection. The components of the unit tangent vector to ℓ at that point are

$$\dot{\xi} = \frac{d}{ds}\,\xi(s) = \tanh s,$$

$$\dot{\eta} = \frac{d}{ds}\,\eta(s) = (\cosh s)^{-2}.$$

The components of the unit tangent vector to the line $\eta = \eta_1$ at that point are

$$\dot{\xi}' = 1, \qquad \dot{\eta}' = 0.$$

According to (5.5-5) and (5.5-6), the angle α is given by

$$\cos\alpha = (\dot{\xi}\ \dot{\eta}) \begin{pmatrix} 1 & 0 \\ 0 & e^{2\xi} \end{pmatrix} \begin{pmatrix} \dot{\xi}' \\ \dot{\eta}' \end{pmatrix} = \tanh s;$$

hence the angle of parallelism satisfies the equation $\cos \Pi(s) = \tanh s$. In the notation of Section 3.1, where the distance s was called y, the formula can be written in any of the following forms:

$$\cos \Pi(y) = \tanh y, \qquad \tan\frac{\Pi(y)}{2} = e^{-y},$$

$$\cot \Pi(y) = \sinh y, \qquad \sin \Pi(y) = \frac{1}{\cosh y}.$$

$$(6.1\text{-}1)$$

In all of these, the unit of length has been so chosen that the constant K in (3.9-3) equals 1 (since that was assumed in Section 5.4). Otherwise, y must be replaced by Ky in the right sides of the above equations.

The equation of a horocycle in polar coordinates r, θ was given in Section 3.9 by Equation (3.9-1) in terms of the function $\Pi(y)$, the formula for which was not known at that time. From (6.1-1), we have explicitly

$$\tanh \frac{r}{2} = \cos \theta. \tag{6.1-2}$$

This horocycle passes through the origin of polar coordinates, where it is orthogonal to the θ-axis. The equation of a general horocycle can be obtained from this by means of the isometries described in Section 6.6 below. As noted in Chapter 3, all horocycles are congruent (there are no "large" and "small" ones), and each horocycle is invariant under an ideal rotation in which the lines of the asymptotic bundle perpendicular to it are mapped onto each other; the horocycle "slides along itself" under the ideal rotation.

6.2 Differential Equations and Formulas for $g(r)$ and $f(r)$

We recall that the function $\phi(a, b, \gamma)$ gives the length c of the third side of a triangle with adjacent sides of lengths a and b and included angle γ (Axiom 6 and Section 2.5) and that the function $\delta(a, b, \gamma)$ gives the angular defect of that same triangle (Section 3.15). In Section 3.15 those functions were considered for thin isosceles triangles, by setting $a = b = r$ and letting $\gamma = \alpha \to 0$. It was found that

$$\lim_{\alpha \to 0} \frac{\phi(r, r, \alpha)}{\alpha} = g(r), \qquad \lim_{\alpha \to 0} \frac{\delta(r, r, \alpha)}{\alpha} = f(r),$$

where $g(r)$ and $f(r)$ were defined by saying that $g(r)\alpha$ is the length of a circular arc of radius r and angle α and $f(r)\alpha$ is the area of a circular sector of radius r, angle α.

Since an isosceles triangle of radius r and angle γ is the union of two right triangles of hypotenuse r and angle $\alpha = \gamma/2$ as in Fig. 6.2a, we can also write

$$g(r) = \lim_{\alpha \to 0} \frac{\sigma(r, \alpha)}{\alpha}, \qquad f(r) = \lim_{\alpha \to 0} \frac{\delta(r, \alpha)}{\alpha},$$

where $\sigma(r, \alpha)$ and $\delta(r, \alpha)$ refer to a right triangle with hypotenuse r and one acute angle α; $\sigma(r, \alpha)$ is the length of the side a opposite the angle α, and $\delta(r, \alpha)$ is the angular defect of that triangle as in Exercise 3 of Section 2.5.

Here we find differential equations for $g(r)$ and $f(r)$; then, by solving those equations, we find the actual formulas. First, for fixed r and Δr, we compute $\sigma(r + \Delta r, \alpha) - \sigma(r, \alpha)$; we then divide by α and let $\alpha \to 0$, to give $g(r + \Delta r) - g(r)$. To first order of small quantities, as $\alpha \to 0$, the angle α'

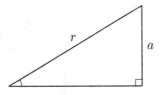

Fig. 6.2a.

in Fig. 6.2b is equal to $\alpha + \delta(r, \alpha) \approx \alpha(1 + f(r))$, $|BC| \approx |DC'|$ (by Section 3.14), and $\sigma(r + \Delta r, \alpha) - \sigma(r, \alpha) \approx |B'D| = \sigma(\Delta r, \alpha') \approx g(\Delta r)\alpha'$. Hence, on dividing by α and Δr and letting $\alpha \to 0$, we have

$$\frac{g(r + \Delta r) - g(r)}{\Delta r} \approx [1 + f(r)]\frac{g(\Delta r)}{\Delta r}.$$

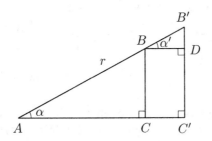

Fig. 6.2b.

Now, $g(r)$ is an increasing function. According to a theorem of Lebesgue,[1] such a function has a (finite) derivative for almost all values of r. If r is one of those values, the left member above has a limit as $\Delta r \to 0$. Therefore, $g(\Delta r)/\Delta r$ has a limit, which we call k_1 (from which it follows that $g(r)$ has a derivative for *all* r), and we have the differential equation

$$g'(r) = k_1[1 + f(r)]. \tag{6.2-1}$$

We now find $f'(r)$. $\delta(r+\Delta r, \alpha) - \delta(r, \alpha)$ is the defect of the quadrilateral $BCC'B'$ in Fig. 6.2c, which is made up of three right triangles, as shown. To first order of small quantities, as $\alpha \to 0$, we have $|AC| \approx r$, $|CC'| \approx \Delta r$, $|BC| \approx g(r)\alpha \approx g(\Delta r)\beta$, so that $\beta \approx (g(r)\alpha)/(g(\Delta r))$, $\beta' \approx \beta$, hence

---

[1] See F. Riesz, B. Sz. Nagy, *Functional Analysis*, Frederick Ungar, New York, 1955, pages 5–9.

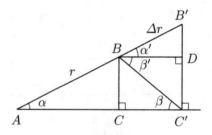

Fig. 6.2c.

$$\delta(r + \Delta r, \alpha) - \delta(r, \alpha) = \operatorname{def}(BB'D) + \operatorname{def}(BDC') + \operatorname{def}(BCC')$$
$$\approx f(\Delta r)[\alpha' + \beta' + \beta]$$
$$\approx f(\Delta r)\alpha \left[1 + 2\frac{g(r)}{g(\Delta r)}\right].$$

Therefore, on dividing by α and Δr and letting $\alpha \to 0$, we have

$$\frac{f(r + \Delta r) - f(r)}{\Delta r} \approx \frac{f(\Delta r)}{\Delta r}\left[1 + 2\frac{g(r)}{g(\Delta r)}\right]$$
$$\approx \frac{f(\Delta r)}{(\Delta r)^2}\left[\Delta r + 2g(r)\frac{\Delta r}{g(\Delta r)}\right].$$

By Lebesgue's theorem, again, the left member has a limit as $\Delta r \to 0$ for almost all r; let r be such a value; since $(\Delta r)/(g(\Delta r))$ has the limit $1/k_1$, as $\Delta r \to 0$, it follows that $f(\Delta r)/(\Delta r)^2$ has a limit, say k_2, so that

$$f'(r) = k_2 2g(r)\frac{1}{k_1} = k_3 g(r). \qquad (6.2\text{-}2)$$

We now consider initial conditions for the differential equations (6.2-1) and (6.2-2). From Section 3.16 we see that $g(0) = f(0) = 0$. Furthermore, from (4.9-3) we see that, since $g(r)$ is the limit of $(\sigma(r, \alpha))/\alpha \approx (r \sin \alpha)/\alpha$, $g(r) = r + \mathcal{O}(r^3)$, as $r \to 0$. From that it follows that the constant k_1 in (6.2-1) is $= 1$. Finally, then, the solutions of the differential equations are

$$g(r) = \frac{1}{K}\sinh Kr, \qquad (6.2\text{-}3)$$
$$f(r) = \cosh Kr - 1, \qquad (6.2\text{-}4)$$

where $K = k_3$.

The constant K, which has the dimensions of an inverse length, will appear in all the formulas in this chapter. As pointed out in Section 1.2, we can take $K = 1$ by a suitable choice of the unit of length in the hyperbolic plane.

It will be seen in the next section that K is the same as the constant K in Equation (3.9-3).

Corollary: In units such that $K = 1$, the circumference and area of a circle of radius r are given by

$$\text{Circumference} = 2\pi \sinh r, \tag{6.2-5}$$

$$\text{Area} = 2\pi(\cosh r - 1). \tag{6.2-6}$$

Exercises

1. Give a geometric interpretation of the values of $g(r)$ and $f(r)$ for very small (but not zero) values of r.
2. Derive formulas for functions corresponding to $g(r)$ and $f(r)$ on the surface of a sphere.

6.3 Formulas for Triangles and Equidistants

We first derive the exact formulas for right triangles (not assumed small). They will be formulas for the legs of the triangle in terms of its hypotenuse and one of the acute angles. We use polar coordinates r, α about an origin A, as in Fig. 6.3a, and we keep b fixed, but let α vary. Then, r, β, a are functions of α. We shall see that they have derivatives with respect to α, which we denote by $(dr)/(d\alpha)$, $(d\beta)/(d\alpha)$, and $(da)/(d\alpha)$. The angular defect of the triangle ABB' is equal to $-\Delta\alpha - \Delta\beta$ (note that $\Delta\beta$ is negative). If we neglect the defect of the small triangle at the top, which is of order $(\Delta\alpha)^2$, that defect is the same as that of the isosceles triangle ABD, which is $2\delta(r, (\Delta\alpha)/2)$ and is approximately equal to $f(r)\Delta\alpha$ according to Section 6.2. Hence $\Delta\beta \approx -\Delta\alpha - f(r)\Delta\alpha$. Dividing by $\Delta\alpha$ and taking the limit gives

$$\frac{d\beta}{d\alpha} = -1 - f(r). \tag{6.3-1}$$

From the approximate formula for the small triangle BDB', where $\angle D$ is approximately a right angle, we have

$$\Delta r \approx |BB'| \cos\beta = \Delta a \cos\beta$$

from which

$$\frac{dr}{d\alpha} = \frac{da}{d\alpha}\cos\beta. \tag{6.3-2}$$

Similarly, $\Delta a \sin\beta \approx |DB| \approx g(r)\Delta\alpha$, from which

$$\frac{da}{d\alpha}\sin\beta = g(r). \tag{6.3-3}$$

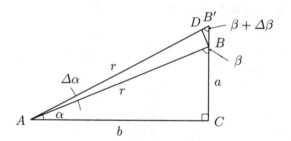

Fig. 6.3a.

From the last two equations (6.3-2) and (6.3-3), we find $(1/g(r))(dr/d\alpha) = \cot\beta$; multiplying through by (6.3-1) gives

$$\frac{-1-f(r)}{g(r)}\frac{dr}{d\alpha} = \cot\beta\frac{d\beta}{d\alpha}.$$

Putting in the values of $g(r)$ and $f(r)$ from (6.2-3) and (6.2-4) and integrating gives

$$-\int K\coth Kr\,dr = \int \cot\beta\,d\beta,$$

hence

$$-\ln\sinh Kr = \ln\sin\beta + \text{const}.$$

By considering the special case where $\alpha = 0$, so that $r = b$, $\beta = (\pi/2)$, and hence $\ln\sin\beta = 0$, we see that the constant is $-\ln\sinh Kb$; therefore, $\sinh Kb = \sinh Kr\sin\beta$. To put this in more standard notation, we interchange a and b and also α and β, and we set $r = c$ for the hypotenuse. That gives

$$\sinh Ka = \sinh Kc\sin\alpha. \qquad (6.3\text{-}4a)$$

The corresponding formula for the side b *adjacent* to the angle α is

$$\tanh Kb = \tanh Kc\cos\alpha; \qquad (6.3\text{-}4b)$$

its derivation is left to Exercise 1 below.

We now consider a general triangle, where a and b are the lengths of two sides and γ is the included angle, as in Fig. 6.3b. The length c of the side opposite the angle γ is given by the function $\phi(a, b, \gamma)$ determined by Axiom 6, and we wish to find the actual formula. The derivation is based on dropping a perpendicular from the vertex at the angle γ to the side c, thus representing the triangle either as the sum or the difference of two right triangles, depending on whether the foot of that perpendicular lies inside the side c or on an extension of that side. The derivation is left to Exercise 2; it gives the formula

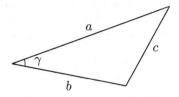

Fig. 6.3b.

$$\cosh Kc = \cosh Ka \cosh Kb - \sinh Ka \sinh Kb \cos\gamma. \qquad (6.3\text{-}5)$$

In the limit of very small a and b, it gives the familiar law of cosines:

$$c^2 = a^2 + b^2 - 2ab\cos\gamma.$$

For $\gamma = (\pi/2)$ Equation (6.3-5) gives the generalization of the Pythagorean theorem, which says that if c is the hypotenuse of a right triangle and a, b are the other sides, then

$$\cosh Kc = \cosh Ka \cosh Kb. \qquad (6.3\text{-}6)$$

By expanding the hyperbolic functions we see that this reduces to the Pythagorean formula for the Euclidean case in the limit of small triangles.

The equation of an *equidistant* to the θ axis is now expressible in terms of the right triangle function. We let r, θ vary in such a way that the distance d in Fig. 6.3c is held fixed. Since $\sinh Kd = \sinh Kr \sin\theta$, we have $\sinh Kr = D/(\sin\theta)$, where $D = \sinh Kd$. More generally, for an equidistant to a line through the origin at angle θ_0, we have

$$\sinh Kr = \frac{D}{\sin(\theta - \theta_0)} \qquad (D, \theta_0 \text{ constants}) \qquad (6.3\text{-}7)$$

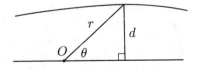

Fig. 6.3c.

This may be contrasted with the equation of a straight line (6.4-1) below. [The change from sine to cosine is irrelevant, but the change from sinh to tanh indicates that the equidistant is not a straight line (except for $d = 0$), but a curve in the hyperbolic plane.]

Exercises

1. Show that for fixed b in Fig. 6.3a, the function $F(\alpha) = \cosh b \cos \alpha - \cosh r \sin \beta$ satisfies a second-order differential equation. From that equation and the initial values at $\alpha = 0$, show that $F(\alpha)$ is identically zero. From that derive Equation (6.3-4b).

2. Derive Equation (6.3-5) by the method described in the text. Show that it gives the law of cosines in the limit of very small a, b, c. Show that if the hyperbolic functions in it are replaced by the corresponding trigonometric ones, the result is a standard equation of spherical trigonometry.

3. Give an alternative derivation of Equation (6.1-1) for the angle of parallelism as follows: In the triangle in Fig. 6.3d, the limiting value of the angle α is $\Pi(y)$, as the distance $|AB|$ goes to infinity, for fixed y.

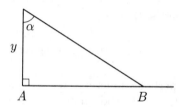

Fig. 6.3d.

4. Find a third derivation for the angle of parallelism as follows: In Fig. 6.3e, ℓ' and ℓ'' are two lines asymptotic to line ℓ. From the equation for the small right triangle at the top of the diagram, find a differential equation satisfied by the function $\Pi(y)$.

5. Recall from Section 3.12 that the area of a triangle is proportional to its angular defect δ. From Exercise 1 and equations (6.3-4a,b) show that in the limit as a, b, δ all tend to zero for fixed angle α, $(ab)/2 = \delta/K^2$. Conclude that if the constant of length is so chosen that $K = 1$ and the constant of proportionality between area and defect is so chosen that the area is equal in the limit to $\frac{1}{2}$ base times altitude, then the area of any triangle (or any polygon) is *equal* to its angular defect.

6. Show that the constant K of this section is the same as the constant K of equation (3.9-3).

7. Deduce the *generalized law of sines* (discovered by Lobachevski) for a general triangle

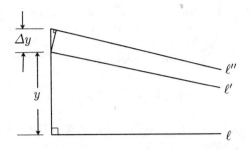

Fig. 6.3e.

$$\frac{\sinh Ka}{\sin \alpha} = \frac{\sinh Kb}{\sin \beta} = \frac{\sinh Kc}{\sin \gamma},$$

where a, b, c are the sides opposite the angles α, β, γ, respectively. *Hint:* Two cases have to be considered.

For the next two exercises, assume the unit of length is chosen so that K is equal to 1.

8. An equilateral triangle has each angle equal to 45 degrees. What is its perimeter?

9. A quadrilateral has each of its angles equal to 60 degrees. What are the possible values of its perimeter?

6.4 Equation of a Line in Polar Coordinates

We consider first a line ℓ that intersects the axis of $\theta = 0$ perpendicularly at a positive distance b from the origin A, as shown in Fig. 6.4. If B is

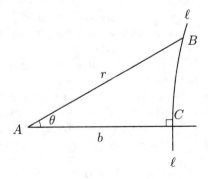

Fig. 6.4.

a variable point on ℓ, with coordinates r, θ, then, from Equation (6.3-4b) for the triangle ABC, we have $\tanh Kb = \tanh Kr \cos \theta$. We denote the constant $\tanh Kb$ by B; then, $\tanh Kr = B/(\cos \theta)$. More generally, if the point on ℓ closest to the origin has polar coordinates b, θ_0, the equation of the line is

$$\tanh Kr = \frac{B}{\cos(\theta - \theta_0)},\tag{6.4-1}$$

where $B = \tanh Kb$; $0 \le B < 1$. For $B = 0$, the line consists of the two rays $\theta = \theta_0 \pm (\pi/2 \ (\text{mod} \ \pi))$, $r \ge 0$.

6.5 The Ideal Points at Infinity

As $r \to \infty$ in (6.4-1),

$$\theta \to \theta_0 \pm \arccos B.$$

Since $-\pi < \theta_0 \le \pi$ and $0 < \arccos B \le (\pi/2)$, we see that the limiting values of θ can be any two distinct values in $(-\pi, \pi] \ (\text{mod} \ 2\pi)$. Hence, any two ideal points at ∞ determine a unique line, as already shown in Section 3.4. The same is true of one ideal point and one finite point (proof left to the reader).

Conclusion: In the hyperbolic plane, any two points (finite or ideal) determine a unique line.

6.6 Formulas for Isometries in Polar Coordinates

In Section 3.6 it was shown that any isometry σ of the hyperbolic plane can be expressed as a finite composition of mappings, each of which is either a rotation R_α about the origin by an angle α, a reflection M in the line ℓ of $\theta = 0$, or a translation T_X by a positive distance X along the line ℓ. In each case, we suppose that a point P with coordinates r, θ is mapped onto a point $Q = \sigma P$ with coordinates r_1, θ_1.

For $\sigma = R_\alpha$, we have $r_1 = r$, $\theta_1 = \theta + \alpha \ (\text{mod} \ 2\pi)$.

For $\sigma = M$, we have $r_1 = r$ and $\theta_1 = -\theta$, except for $\theta = \pi$, in which case we have $\theta_1 = \pi$. (Recall that we defined θ so that $-\pi < \theta \le \pi$.)

We now consider translations. Under T_X, a point P with coordinates r, θ is carried to a point $Q = \sigma P$ with coordinates r_1, θ_1 as shown in Fig. 6.6; we have, by use of the formulas for right triangles, first[2]

$$\sinh r_1 \sin \theta_1 = \sinh r \sin \theta,\tag{6.6-1}$$

---

[2] In this section, we assume the unit of length so chosen that the constant K, which appears in the preceding sections, is equal to 1 — see Section 1.2.

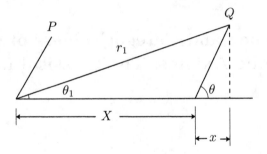

Fig. 6.6.

which says that P and Q are both at the same distance from the axis $\theta = 0$, and

$$\tanh r_1 \cos \theta_1 = \tanh(X + x),$$
$$\tanh r \cos \theta = \tanh x,$$

which say that the perpendiculars dropped from P and Q to the axis $\theta = 0$ are separated along that axis by distance X. By the formula for the hyperbolic tangent of a sum, the last two equations can be replaced by

$$\tanh r_1 \cos \theta_1 = \frac{\tanh X + \tanh r \cos \theta}{1 + \tanh X \tanh r \cos \theta}. \qquad (6.6\text{-}2)$$

In the next chapter, Equations (6.6-1) and (6.6-2) will be put into a simpler form by use of the coordinates of the Poincaré disk model.

Chapter 7

Consistency and Categoricalness of the Hyperbolic Axioms; The Classical Models

We take as primary model of the hyperbolic plane an abstract surface $\mathcal{S}$ in the sense of Section 5.3, whose geometry is determined by the methods of differential geometry in such a way that all the axioms of the hyperbolic plane are satisfied. We consider several different coordinate systems, each of which covers the entire surface $\mathcal{S}$, some of which are more useful than others for certain purposes. Each coordinate system leads to one of the classical models of the hyperbolic plane based on Euclidean geometry. Differential geometry is based on analysis, which is based on the real number system $\mathbb{R}$. It follows that the hyperbolic axioms are consistent, if the axioms of $\mathbb{R}$ are consistent. It is proved that the axiom system is categorical, in the sense that any model of the hyperbolic plane is isomorphic to any other model. Lastly, as an amusement, we describe a hyperbolic model of the Euclidean plane.

7.1 Models

The nature and purpose of models were discussed in the third section of Chapter 3. In this chapter, we present models in which "points," "lines," "distances," and "angles" are defined as certain things in a structure based on earlier parts of mathematics, namely, on analysis, or on Euclidean geometry, or on both. If we can prove that the relations that exist among these "points," "lines," "distances," and "angles" satisfy all the axioms of the hyperbolic plane, we have a model, and the existence of the model shows that the hyperbolic axioms are consistent (i.e., cannot lead to a contradiction), on the assumption (which everyone believes) that the earlier axioms are consistent, because if the hyperbolic axioms could lead to a contradiciton, then the same contradiction would appear in the model as a consequence of the earlier axioms.

Each is an abstract surface $\mathcal{S}$ whose geometry is specified by giving a line element or metric tensor in a region Ω of parameter space; the parameters u, v are thought of as "coordinates" in the surface. Then the "points" are the points of $\mathcal{S}$, the "lines" are the geodesics, and the "distance" between two points P and Q is the arclength between P and Q along the geodesic that joins them; the "angles" are the angles determined by the line element as in Section 5.5.

We now consider various coordinate systems in $\mathcal{S}$ and the corresponding line elements.

7.2 Definition of $\mathcal{S}$ and the Coordinates ξ, η

We use the coordinates ξ, η defined in Section 3.9, and we take the region Ω to be the entire ξ, η plane. According to that section, we then have a one-to-one mapping between the abstract surface $\mathcal{S}$ and the hyperbolic plane $\mathbb{H}^2$, and we take the geometry of $\mathcal{S}$ to be defined by the line element and metric tensor given in Section 5.3, namely,

$$ds^2 = d\xi^2 + e^{2\xi} d\eta^2, \qquad G = \begin{pmatrix} 1 & 0 \\ 0 & e^{2\xi} \end{pmatrix}. \qquad (7.2\text{-}1)$$

Note: The mapping between $\mathcal{S}$ and $\mathbb{H}^2$ could have been established by any of the other coordinate systems discussed below, each of which covers the entire plane. When the coordinates are changed, we think of the points of $\mathcal{S}$ as remaining fixed; it is simply the number pairs (u, v) that change.

7.3 The Poincaré Half-Plane Coordinates

The half-plane coordinates x, y are defined in terms of ξ, η as

$$x = \eta, \quad y = e^{-\xi}, \quad z = x + iy. \qquad (7.3\text{-}1)$$

The region Ω is now the half-plane $(y > 0,$ all $x)$. By transforming the line element of the preceding section to the new coordinates, we find

$$ds^2 = \frac{dx^2 + dy^2}{y^2}, \qquad G = \left(\frac{1}{y}\right)^2 \begin{pmatrix} 1 & 0 \\ 0 & 1 \end{pmatrix}. \qquad (7.3\text{-}2)$$

Since the metric tensor is a multiple of the unit matrix, these coordinates are conformal: The angle of intersection of any two curves in $\mathcal{S}$, as defined in Section 5.5, is equal to the angle of intersection of the corresponding curves in Ω.

The "lines" of the model are the geodesics in $\mathcal{S}$. They are given in the ξ, η coordinates by (5.4-9) and (5.4-10), hence by

$$\left\{ \begin{array}{l} x - \eta_0 = e^{-\xi_0} \tanh s \\[2mm] y = e^{-\xi_0} \dfrac{1}{\cosh s} \end{array} \right\} \quad \text{and} \quad \left\{ \begin{array}{l} x = \eta_0 \\ y = e^{-s} \end{array} \right\}. \qquad (7.3\text{-}3)$$

where ξ_0 and η_0 are arbitrary real constants. As curves in Ω, they are

$$(x - \eta_0)^2 + y^2 = (e^{-\xi_0})^2 \quad \text{and} \quad x = \eta_0, \text{ all } y > 0, \tag{7.3-4}$$

that is, semicircles centered on the x-axis and vertical half-lines. The direction of increase of the arclength parameter is as indicated in Fig. 7.3; it is of course arbitrary. The value of s could be replaced by $-s$ in either or both of (7.3-3).

Fig. 7.3. Geodesics

Before showing that the model satisfies all the hyperbolic axioms, we first discuss its isometries.

7.4 The Isometries of $\mathcal{S}$ in Terms of the Half-Plane Coordinates

The isometries of the model are the one-to-one mappings $x, y \to x', y'$ of the upper half-plane onto itself under which the metric (7.3-2) is preserved, that is, under which

$$\frac{dx^2 + dy^2}{y^2} = \frac{dx'^2 + dy'^2}{y'^2}. \tag{7.4-1}$$

That the mapping is one-to-one and onto means that it satisfies the following requirement:

Requirement: Any point x, y with $y > 0$ is mapped onto a unique point x', y' with $y' > 0$, and, conversely, any point x', y' with $y' > 0$ is the image under the mapping of a unique point x, y with $y > 0$.

First, it is clear that the requirement is satisfied and the metric is preserved under a mapping in which we merely add a constant to x or under which we multiply both x and y by the same positiive constant. In complex notation, with $z = x + iy$, we have mappings of the form $z \to z' = az + b$, where a and b are real, and a is positive. They are direct isometries and they represent the isometries of the two-parameter subgroup discussed in terms of the ξ, η coordinates in Section 3.9. A reverse isometry (reflection) is given by $z' = -\bar{z}$, where the overbar denotes complex conjugate. We now show that the metric is also preserved by the more general fractional-linear transformation

$$z \to z' = \frac{az + b}{cz + d},\tag{7.4-2}$$

where a, b, c, d are real and $ad - bc$ is positive (these conditions are necessary for satisfying the above requirement). Since the above mapping is unaltered if we multiply a, b, c, d all by a nonzero constant, we can require that $ad - bc$ be equal to 1. One way to show the invariance of the metric is to use complex notation, so that $dx^2 + dy^2 = (dx + idy)(dx - idy) = dz d\bar{z}$. By differentiating (7.4-2) and using $ad - bc = 1$, we find that

$$dz' = \frac{dz}{(cz + d)^2}$$

and

$$2iy' = z' - \bar{z}' = \frac{z - \bar{z}}{(cz + d)(c\bar{z} + d)},$$

from which

$$\frac{dz' d\bar{z}'}{(z' - \bar{z}')^2} = \frac{dz d\bar{z}}{(z - \bar{z})^2}.$$

This is identical with (7.4-1) except for a factor $-(1/4)$ on each side of the equation.

Some of the new isometries that we have found in this way (when $c \neq 0$) differ from the purely linear ones $z \to z' = az + b$ in that they keep fixed a point in the upper half-plane, hence represent rotations in the hyperbolic plane. It is easily seen that if Z and Z' are any points in the half-plane and $\vec{k}$ and $\vec{k}'$ are any rays drawn from those points (along geodesics), there is a choice of the constants a, b, c, d so that the transformation takes Z onto Z' and $\vec{k}$ onto $\vec{k}'$. Therefore, according to Sections 3.6 and 3.8, the set of all mappings (7.4-2) with a, b, c, d real and $ad - bc = 1$ represent the entire group of direct isometries in the hyperbolic plane. (For direct isometries, it is not necessary to specify the choices of the positive half-planes HP and HP'.)

The isometries will be discussed in more detail in the next chapter.

7.5 The Poincaré Half-Plane Satisfies the Hyperbolic Axioms

That any two points determine a unique "line" (geodesic) is clear. If one of the points in the half-plane Ω lies vertically above the other, the unique geodesic is the vertical half-line that passes through them. Otherwise, the center of the unique semicircle is found by extending the perpendicular

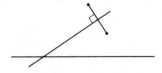

Fig. 7.5a.

bisector of the segment joining the points until it intersects the x-axis, as in Fig. 7.5a.

The distance function $|PQ|$ of Axiom 2 is the arclength from P to Q along the unique geodesic that joins them. To verify the triangle inequality (1.1-1), we first show that the hypotenuse of a right triangle is always longer than either leg (in this model, that is; it is not true on a sphere). By one of the isometries (7.4-2), move the triangle, if necessary, so that one of its legs is vertical in the half-plane, with the right angle at its lower end, as in Fig. 7.5b.

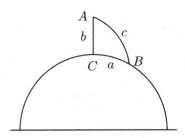

Fig. 7.5b.

We denote the coordinates of the point A by x_A, y_A, and similarly for the other vertices. We denote the lengths of the sides (arclength along geodesics) by a, b, c, as shown. From the second pair of formulas in (7.3-3), the arclength along the vertical geodesic is $s = -\ln y$, hence $y_A/y_C = e^b$. From the first pair of formulas for the semicircle through A and B, we have that y_A/y_B is equal to $\cosh(s_A + c)/\cosh s_A = \cosh c + \tanh s_A \sinh c < \cosh c + \sinh c = e^c$. Since clearly $y_A/y_C < y_A/y_B$, we see that $b < c$, as was to be proved. The triangle inequality for a general triangle is then established by dropping a perpendicular from one vertex to the opposite side (extended, if necessary) and applying the above law to the right triangles so formed. Hence Axiom 2 is satisfied by the model.

Axiom 3 is satisfied by taking the coordinate x to be the arclength s. The formulas (7.3-3) show that s varies from $-\infty$ to $+\infty$ on each geodesic.

Axiom 4 is evidently satisfied; each vertical half-line or semicircle centered on the x-axis separates the upper half-plane ($y > 0$) into two subsets as required.

It was pointed out in Section 5.5 that angles, as defined there in terms of the metric, satisfy the requirements of Axiom 5.

Axiom 6 (the SAS criterion for triangles) follows from the isometries (fractional linear transformations) of the model. Given any two triangles that have two congruent sides and congruent included angle, there exists an isometry that moves the vertex (where the included angle lies) of the first triangle to the corresponding vertex of the second and moves the two sides of the first triangle into coincidence with the corresponding sides of the second triangle. Since the "moving" was done by an isometry, it is now seen that the two triangles are fully congruent.

Lastly, it is evident from Fig. 7.5c that given a geodesic ℓ and a point P not on it, there are many geodesics through the point P that do not intersect the first geodesic anywhere in the half-plane; hence Axiom 7b is satisfied.

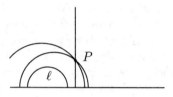

Fig. 7.5c.

7.6 Consistency of the Hyperbolic Axioms

As noted in Section 7.1 above, the existence of a model proves the consistency of the hyperbolic axioms, on the assumption that the axioms of real analysis are consistent.

7.7 Categoricalness of the Hyperbolic Axioms

We shall show that any two models of the hyperbolic plane are isomorphic. The proof is based on polar coordinates r, θ, which are defined from the axioms, as in Section 3.3, once one has chosen the following:

(1) an origin Z;
(2) a direction from that origin as the axis where $\theta = 0$;

(3)   one of the half-planes bounded by that axis as the half-plane where
$0 < \theta < \pi$; and

(4)   a unit of length.

We shall always take the unit of length so that $K = 1$, where K is the constant that appeared in the preceding chapter. (See Section 1.2 for comments on the unit of length.)

Given two models, $\mathcal{M}$ and $\mathcal{M}'$, we assume that the choices (1), (2), and (3) have been made in each model, and we define a mapping σ from $\mathcal{M}$ to $\mathcal{M}'$ by saying that if r, θ are the coordinates of a point P in $\mathcal{M}$, then its image $\sigma(P)$ in $\mathcal{M}'$ is the point of $\mathcal{M}'$ with coordinates $r' = r$, $\theta' = \theta$. Clearly, σ is a one-to-one mapping of $\mathcal{M}$ onto all of $\mathcal{M}'$, and we shall show that it preserves all geometrical relations. Among the things that have to be proved are:

(a) If three points P, Q, R in $\mathcal{M}$ are collinear, then the images P', Q', R' in $\mathcal{M}'$ are also collinear. That is, if ℓ is the line in $\mathcal{M}$ determined by the points P, Q, and ℓ' the line determined by P', Q', then R lies in ℓ if and only if R' lies in ℓ'. The proof is given below.

(b) Distances and angles are preserved by σ. In particular, we must show that if $|PQ|$ denotes the distance function in $\mathcal{M}$ and $|RS|'$ denotes that in $\mathcal{M}'$, then $|PQ| = |\sigma(P)\sigma(Q)|'$ for all pairs P, Q of points. From the definition of polar coordinates, it is seen that the distance from the origin Z to any point P, which is by definition r, is the same as the distance from $\sigma(Z)$ to $\sigma(P)$, because we have set $r' = r$. What has to be proved is that distances between general points P, Q are also preserved. The distance between points with polar coordinates r_1, θ_1 and r_2, θ_2 is given in terms of the triangle function $\phi(a, b, \gamma)$ (which came from Axiom 6) as $\phi(r_1, r_2, \theta_1 - \theta_2)$. A formula for the function ϕ is given in Equation (6.3-5), derived directly from the axioms, in a way not dependent on any model, and since the polar coordinates r, θ are preserved, it follows that the distance from r_1, θ_1 to r_2, θ_2 is preserved, since the models $\mathcal{M}$ and $\mathcal{M}'$ satisfy Axiom 6.

The preservation of collinearity mentioned in (a) above then follows from a neutral theorem which says that a point P lies in a line ℓ if and only if the greatest lower bound of the distances $|PX|$, as X varies in the line, is zero. Hence, lines are mapped onto lines.

It is known (see Theorem 6 in Section 2.4 and Section 4.6) that the function $\phi(a, b, \gamma)$ is a continuous increasing function of γ for fixed a, b. Therefore the equation $c = \phi(a, b, \gamma)$ has an inverse $\gamma = \psi(a, b, c)$, which is also the same for all models, and which determines any angle $\angle ABC$ in terms of distances. It follows that all angles are invariant under the mapping σ.

Since all geometric properties can be expressed in terms of distances, angles, and lines, we have proved the main theorem:

Theorem 1: The mapping σ defined above is an isomorphism of the two models. Therefore, the axioms of hyperbolic plane geometry are categorical.

Additional models are discussed in the next section. Henceforth, any of them may be used to draw general conclusions about hyperbolic geometry.

There is here an apparent conflict between categoricalness and Gödel incompleteness similar to the one discussed in Section A.6 of the Appendix to Chapter 1.

Owing to the necessity of making the choices (1), (2), and (3) above for each model, the mapping σ is highly nonunique. That reflects the homogeneity and isotropy of the hyperbolic plane that results from the isometries. No matter where you stand in the plane, and no matter in which direction you look, the scenery is exactly the same as if you stand at another point and look in another direction. The same is true of any model.

The isomorphism can often be expressed in terms of coordinates other than polar. In the next section, the isomorphism between the Poincaré half-plane model and the Poincaré disk model will be established by use of Cartesian coordinates (more precisely, of the resulting complex coordinates). That will prove that the disk model really *is* a model, that is, satisfies all the hyperbolic axioms, because the half-plane model satisfies them all, according to Section 7.5 above.

7.8 Other Models

Classically, the models of the hyperbolic plane were regarded as based on Euclidean geometry. One started with a piece of a Euclidean plane — a half-plane or a circular disk and defined in it "points," "lines," "distances," "angles" as things that could be described in that half-plane or disk in terms of Euclidean geometry. We prefer to think of the models as based on analysis, via differential geometry. We start with an abstract surface S, as in Section 5.3, with its geometry determined by a line element or, equivalently, a metric tensor, in terms of parameters u, v.

The difference is hardly more than a change of point of view. Whether one thinks of the u, v-plane as a Euclidean plane in which u, v are Cartesian coordinates is unimportant. One can think of that plane as simply the Cartesian product $\mathbb{R} \times \mathbb{R}$ of the real number system $\mathbb{R}$ with itself, so that an element of that product is a pair of real numbers u, v. We shall make no use of Euclidean geometry in the u, v-plane. It is in any case a matter of great efficiency to be able to define a model by giving just the range Ω of variation of the parameters u, v and the line element, from which the entire geometry of the model follows.

Each of the coordinate systems considered earlier provides a model. In each case the coordinates cover the entire hyperbolic plane, and we shall give explicit formulas for transforming from one set of coordinates to another, which lead to the transformation of the line element from one set to another, thus showing directly that all these models are isomorphic. Therefore, since the half-plane model has been shown to satisfy all the axioms, it follows that the others do too. It was shown in the preceding section that *any* model that satisfies all the axioms is isomorphic to the abstract hyperbolic plane determined by the axioms.

(1) *The model based on the ξ, η coordinates.*

$$\Omega : \begin{pmatrix} -\infty < \xi < \infty \\ -\infty < \eta < \infty \end{pmatrix}, \qquad ds^2 = d\xi^2 + e^{2\xi} d\eta^2.$$

(2) *The Poincaré half-plane model.*

$$\Omega : \begin{pmatrix} 0 < y < \infty \\ -\infty < x < \infty \end{pmatrix}, \qquad ds^2 = \frac{dx^2 + dy^2}{y^2}. \qquad (7.8\text{-}1)$$

The transformation law connecting these two models is $x = \eta$, $y = e^{-\xi}$.

(3) *The polar-coordinate model.* Let r, θ be polar coordinates in the hyperbolic plane as defined in Section 3.4. According to (5.3-3), the line element is $ds^2 = dr^2 + g(r)^2 d\theta^2$; according to Equations (6.2-3), $g(r)$ is given by $\sinh r$, in units such that $K = 1$. Therefore, this model is given by

$$\Omega : \begin{pmatrix} 0 \le r < \infty \\ -\pi < \theta \le \pi \end{pmatrix}, \qquad ds^2 = dr^2 + (\sinh r)^2 d\theta^2. \qquad (7.8\text{-}2)$$

We shall give below the formulas for transforming from this model to the Poincaré disk model and the formulas for transforming from the disk model to the Poincaré half-plane model.

(4) *The Poincaré disk model.* We introduce coordinates u, v defined in terms of r, θ by the equations

$$u = \tanh \frac{r}{2} \cos \theta, \qquad v = \tanh \frac{r}{2} \sin \theta.$$

Clearly, these coordinates vary in the disk $u^2 + v^2 < 1$. A short calculation gives the line element in u, v, and we have

$$\Omega : u^2 + v^2 < 1, \qquad ds^2 = 4\frac{du^2 + dv^2}{(1 - u^2 - v^2)^2}. \qquad (7.8\text{-}3)$$

It can be shown, either by computing the geodesics of this metric or by transforming Equation (6.4-1) to the coordinates u, v, that the lines of $\mathbb{H}^2$ are represented in this model by circular arcs in the disk Ω that meet the bounding circle $u^2 + v^2 = 1$ orthogonally, including the diameters of Ω, which may be regarded as circular arcs of infinite radius. (See Exercises below.)

We recall from Section 3.4 that in terms of a given polar coordinate system, each value of θ represents an ideal point at infinity, so that in the Poincaré disk model, each point on the bounding circle represents an ideal point.

The isometries of this model can be expressed in terms of the complex coordinate $w = u + iv$. First, we have

$$w \to w' = e^{i\alpha}w, \quad (0 \le \alpha < 2\pi), \tag{7.8-4}$$

which represents a rotation in $\mathbb{H}^2$ about the origin. Next, we have

$$w \mapsto w' = \frac{w\cosh(X/2) + \sinh(X/2)}{w\sinh(X/2) + \cosh(X/2)} \tag{7.8-5}$$

According to Exercise 3 below, these mappings map the disk $u^2 + v^2 < 1$ onto itself and leave the line element (7.8-3) invariant, hence they represent isometries of the hyperbolic plane. Since (7.8-5) leaves the real axis $(v = 0)$ invariant, the corresponding isometry of the hyperbolic plane leaves the line of the θ-axis invariant, hence is a translation along that line (Exercise 7).

An isomorphism of this model and the Poincaré half-plane model is given by the equations

$$z = x + iy = i\frac{i+w}{i-w}, \qquad w = u + iv = i\frac{z-i}{z+i}. \tag{7.8-6}$$

The *Beltrami–Klein* model, which was historically the first model of the hyperbolic plane, is based on coordinates given by the equations

$$X = \tanh r \cos\theta, \qquad Y = \tanh r \sin\theta. \tag{7.8-7}$$

Each point of the unit disk $X^2 + Y^2 < 1$ in the X, Y-plane represents a point of the hyperbolic plane $\mathbb{H}^2$. From (6.4-1) we see that the lines of $\mathbb{H}^2$ are represented by line segments (chords of the unit circle) in the model.

The model is not conformal; hence the angle between two lines in $\mathbb{H}^2$ is not in general equal to the angle between the corresponding line segments in the model. An important exception to that is that perpendicular line segments in the model represent perpendicular lines in $\mathbb{H}^2$ if (and only if) one of the lines passes through the origin of the polar coordinate system.

If (7.8-1) is transformed from polar coordinates r, θ to the coordinates X, Y, we find that the line element is given by the equation

$$ds^2 = \frac{1}{(1 - X^2 - Y^2)^2} \left[(1-Y^2)dX^2 + (1-X^2)dY^2 + 2XYdXdY \right]. \tag{7.8-8}$$

Because of the last term in the square brackets, the coordinate system X, Y is not conformal and not even orthogonal.

The coordinates (7.8-7) will play an important role in the theory of construction by straightedge and compass in Chapter 11. The Beltrami–Klein model will be generalized to a spherical model of the three-dimensional hyperbolic space in Chapter 9.

Exercises.

Please read these exercises, to see what they say, even if you don't do them.

1. Show that in the Poincaré half-plane model, the vertical half-lines represent a pencil of lines in the hyperbolic plane asymptotic in the upward direction. Show that the horizontal lines ($y = $ const) represent horocycles.

2. Assume that the point $z = i$ of the half-plane model represents the origin of the hyperbolic plane and that the horizontal "line" through $z = i$ (the unit semicircle) represents the θ-axis. Show that the ideal point $\theta = (\pi/2)$ is represented by the infinite points in the half-plane and the other ideal points are represented by the points on the x-axis according to the equation

$$x = \frac{1 + \sin\theta}{\cos\theta}.$$ (7.8-9)

Hint: From the conformality of the model, show that the angle θ in Fig. 7.8a is equal to the polar angle θ in the hyperbolic plane. What happens if θ is replaced by $\theta + \pi$? What is the interpretation of the above formula for $\theta = (3\pi)/2$?

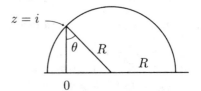

$z = i$

θ R

R

0

Fig. 7.8a.

3. Show that under the mappings (7.8-4) and (7.8-5), the Poincaré disk is mapped onto itself and the metric defined in (7.8-3) is invariant.

4. Show, by either of the methods mentioned in the text, that the geodesics in the disk model are circular arcs orthogonal to the bounding circle $u^2 + v^2 = 1$.

5. Show that under a fractional linear transformation of the complex plane

$$z \to z' = \frac{\alpha z + \beta}{\gamma z + \delta} \qquad (\alpha\beta - \gamma\delta \neq 0),$$

circles are mapped onto circles (where straight lines are included as circles of infinite radius).

6. From Exercise 5 show that a circle in the Poincaré disk model that lies entirely in the disk represents a circle in the hyperbolic plane. *Hint:* This is evident for a circle centered at the origin; then use the mapping (7.8-5), which represents an isometry of the hyperbolic plane.

7. Show that (7.8-5) represents a translation by a distance X in $\mathbb{H}^2$.
8. Describe the trajectory of a point in the disk model (not on the real axis) under (7.8-5) as X varies from $-\infty$ to $+\infty$. What kind of curve does that trajectory represent in the hyperbolic plane?
9. Show that under the mapping (7.8-6) the line element $ds^2 = y^{-2}(dx^2 + dy^2)$ of the half-plane model goes into the line element of the disk model in (7.8-3).
10. Show by another application of Exercise 5 that circles in the half-plane model are mapped by (7.8-6) into circles in the disk model. Conclude that circles that lie entirely in the half-plane represent circles in the hyperbolic plane.
11. How are general horocycles represented in the half-plane model? In the disk model?
12. Find the "distance" function $D(w_1, w_2)$ for the Poincaré disk model. *Hint:* That's easy if $w_1 = 0$; then use isometries. Transform the result to give a "distance" function $D(z_1, z_2)$ for the half-plane model.
13. Interpret Fig. 3.10c and the comment associated with it in terms of the Poincaré disk model.

7.9 The Pseudosphere or Tractroid

Let C be the curve in the r, z-plane as in Fig. 7.9a, given parametrically by

$$r = \frac{1}{\cosh t}, \quad z = t - \tanh t \quad (0 \leq t < \infty).$$

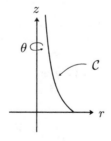

Fig. 7.9a.

The surface obtained by rotating this curve around the z-axis is called a *pseudosphere*. Let θ be an angle about the z-axis, so that r, θ, z are cylindrical coordinates. We investigate the geometry of this surface inherited from the three-dimensional geometry in which it is embedded by the methods of

Section 5.2. The Euclidean element of distance ds in cylindrical coordinates is given by

$$ds^2 = dr^2 + dz^2 + r^2 d\theta^2$$
$$= \left(\frac{\sinh t}{\cosh^2 t} \right)^2 dt^2 + \left(1 - \frac{1}{\cosh^2 t} \right)^2 dt^2 + r^2 d\theta^2$$
$$= \left(\frac{\sinh t}{\cosh t} \right)^2 dt^2 + \frac{1}{\cosh^2 t} \, d\theta^2.$$

If we introduce (curvilinear) coordinates x, y in the surface defined by

$$x = \theta \quad (0 \leq x < 2\pi)$$
$$y = \cosh t \quad (1 \leq y < \infty),$$

we have

$$ds^2 = \frac{dx^2 + dy^2}{y^2}.$$

By comparison with (7.3-2), we see that the intrinsic geometry of the pseudosphere is identical with that of the region of the Poincaré half-plane model shown in Fig. 7.9b. That represents a horocycle sector of the hyperbolic plane, because the vertical lines at $x = 0$, $x = 2\pi$ represent lines asymptotic at infinity at the ideal point at $\theta = (\pi/2)$, and the line segment at $y = 1$ represents an arc of a horocycle perpendicular to those lines.

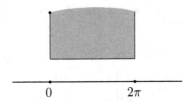

0   2π

Fig. 7.9b.

By extending the region shown in Fig. 7.9b indefinitely horizontally in both directions, we see that the universal covering surface of the pseudosphere (the surface obtained by wrapping a surface infinitely many times around the pseudosphere like paper towels around a cardboard tube) represents the entire interior of the horocycle.

7.10 A Hyperbolic Model of the Euclidean Plane

A colleague has posed the question whether there exists in the hyperbolic plane a model of the Euclidean plane. That reverses the question answered by Klein and Poincaré, when they constructed models of the hyperbolic plane in the Euclidean plane. Such a model can be constructed as follows:

(1) Each point of the hyperbolic plane $\mathbb{H}^2$ is a "point" of the model. We now choose an arbitrary point O of $\mathbb{H}^2$ as origin.

(2) Each line through O is a "line" of the model. Furthermore, any curve equidistant from a line through O is also a "line" of the model. Recall that a curve C is *equidistant* from a line ℓ if the distance d from a point Q of C to ℓ, obtained by dropping a perpendicular from Q to ℓ, is the same for all points Q of C. (Such a curve is not a line in $\mathbb{H}^2$ except in the case $d = 0$.)

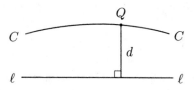

Fig. 7.10a.

Let r, θ be polar coordinates about O in $\mathbb{H}^2$ where r is the distance (i.e., hyperbolic distance) of a point P from O, and θ is the angle between the ray $\overrightarrow{OP}$ and a fixed ray, defined as in Section 3.4. From the formulas of hyperbolic geometry, it is found that the equation of an equidistant to the line $\theta = \theta_0$ is

$$\sinh r = \frac{\sinh a}{\sin(\theta - \theta_0)}, \quad \begin{array}{l} 0 < \theta - \theta_0 < \pi \text{ for } a > 0, \\ -\pi < \theta - \theta_0 < 0 \text{ for } a < 0 \end{array} \qquad (7.10\text{-}1)$$

where a and θ_0 are constants. From this it can be proved that for any two points, there is exactly one "line" that passes through them both. That is of course one of the axioms of Euclidean geometry. Equidistants with $\theta_0 = 0$ are called "horizontal lines" and ones with $\theta_0 = \pi/2$ are called "vertical lines."

(3) Through each point there is just one "vertical line" and just one "horizontal line." Let the equidistants that represent those "lines" be given by

$$\sinh r = \frac{\sinh a}{\cos \theta}; \qquad \sinh r = \frac{\sinh b}{\sin \theta}. \qquad (7.10\text{-}2)$$

Then the numbers x, y given by

$$x = \sinh a \qquad y = \sinh b \qquad\qquad (7.10\text{-}3)$$

are called the "Cartesian coordinates" of the point.

(4) We define the "distance" between points P, P' with "Cartesian coordinates" x, y and x', y' as

$$[(x - x')^2 + (y - y')^2]^{\frac{1}{2}} \stackrel{\text{def}}{=} \rho(P, P'). \qquad\qquad (7.10\text{-}4)$$

In particular, for $x' = y' = 0$, we find that the "distance" of a point from the origin is, according to (2) and (3).

$$\rho(P, 0) = [x^2 + y^2]^{\frac{1}{2}} = [(\sinh a)^2 + (\sinh b)^2]^{\frac{1}{2}} = \sinh r. \qquad (7.10\text{-}5)$$

(5) "Angles" are defined in terms of the "Cartesian coordinates" x, y in the usual way, namely, if $\alpha = \angle PP'P''$, then, by the dot-product formula,

$$(x - x')(x'' - x') + (y - y')(y'' - y') = \rho(P, P')\rho(P'', P') \cos \alpha. \quad (7.10\text{-}6)$$

As a special case, it is easily seen that "angles" about the origin in the model are equal to the corresponding angles about O in $\mathbb{H}^2$.

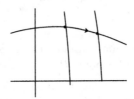

Fig. 7.10b.

(6) We now consider a one-to-one mapping of $\mathbb{H}^2$ onto itself, in which a point with parameters a and b in equation (7.10-2) is moved along the "horizontal" equidistant on which it lies to another point on that equidistant, as in Fig. 7.10b, so that it then lies on a "vertical" equidistant with the parameter a replaced by a', where $\sinh a' - \sinh a = \text{const}$, say A. That corresponds to the mapping $x' = x + A$, $y' = y$ in the model, hence is called a "translation." Clearly, all "distances" and "angles" as defined above are invariant under it.

(7) It is easily seen that a rotation through angle α about O in corresponds to the mapping $x' = x \cos \alpha - y \sin \alpha$, $y' = x \sin \alpha + y \cos \alpha$ in the model. Hence, "distances" and "angles" are invariant under it also. The

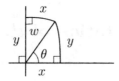

Fig. 7.10c.

group generated by these mappings is an isometry group of the model, and clearly it is isomorphic to the Euclidean isometry group.

(8) If we call $w = \sinh r$, then Equations (7.10-2) and (7.10-3) give

$$x = w \cos \theta, \qquad y = w \sin \theta;$$

hence each of the right triangles in Fig. 7.10c satisfies the Euclidean (trigonometric) formulas. Any general triangle with one vertex at O and the side opposite it "vertical" can be obtained by adding or subtracting right triangles of the kind considered above; hence the Euclidean formulas hold for it. Furthermore, any triangle in the plane can be brought by a first rotation about the origin so that one vertex lies on the line $y = 0$; that vertex can then be brought to the origin by a translation. Then its opposite side can be made vertical by another rotation. Since "lengths" and "angles" are invariant under these operations, it follows that the Euclidean formulas hold for any triangle. That implies the SAS axiom of Euclidean geometry.

(9) It follows from the above that the geometry of the model is either Euclidean or hyperbolic. But it also follows that triangles have zero angular defect. It is a general theorem that then the geometry of the model is Euclidean, and the Euclidean parallel axiom holds.

Chapter 8

Matrix Representation of the
Isometry Group

It is shown that the group of direct isometries of the hyperbolic plane is isomorphic to the group of fractional linear transformations with real coefficients. The coefficients of such a transformation can be taken as the entries of a 2 × 2 real matrix with determinant 1; there is a one-to-two relation between the transformations and such matrices, where a matrix M and its negative $-M$ lead to the same fractional linear transformation, hence to the same isometry of the hyperbolic plane. Those matrices constitute the group $SL(2, \mathbb{R})$ of 2 × 2 real matrices of determinant 1, and it follows that the group of direct isometries of the hyperbolic plane is isomorphic to the quotient group of $SL(2, \mathbb{R})$ with respect to its subgroup consisting of I and $-I$, where I is the 2 × 2 unit matrix. This makes it possible to study the isometries by studying the matrices. The points, lines, and curves that are invariant (mapped onto themselves) under an isometry are classified. It is proved that the direct isometry group of the hyperbolic plane is *simple*, has no nontrivial proper normal subgroups, hence differs in this respect from the isometry group of the Euclidean plane, where the translations form a normal subgroup. Lastly, the subgroup $SL(2, \mathbb{Z})$ of $SL(2, \mathbb{R})$ consisting of matrices with integer entries is studied; it leads to a tiling of the hyperbolic plane by triangles each having two 60° vertices and one 0° vertex at infinity.

8.1 Fractional Linear Transformations

In the preceding chapter it was proved that the axioms of the hyperbolic plane are categorical; any set of objects that satisfies all the axioms has the same geometrical properties as any other such set. In particular, the isometries of a model represent the isometries of the hyperbolic plane. We choose the Poincaré half-plane model. It was shown in Section 7.4 that the line element of that model, given by $ds^2 = y^{-2}(dx^2 + dy^2)$, is invariant under the transformations

$$z \to z' = \frac{az + b}{cz + d}, \tag{8.1-1}$$

where a, b, c, d are real constants, with $ad - bc$ positive, and $z = x + iy$.

It is convenient to collect together the constants in the form of a matrix $\begin{pmatrix} a & b \\ c & d \end{pmatrix}$. Clearly, the transformation is unchanged if the entries are all multiplied by any nonzero constant, and it is convenient to choose that constant so that $ad - bc = 1$; that is, so that the matrices all have determinant 1. Then, there is a two-to-one relation between the matrices and the fractional linear transformations; the matrices $\begin{pmatrix} a & b \\ c & d \end{pmatrix}$ and $\begin{pmatrix} -a & -b \\ -c & d \end{pmatrix}$ give the same fractional linear transformation, hence the same isometry.

It is easily established that if the above transformation is followed by another such, say with constants a', b', c', d', the result is another fractional linear transformation whose matrix is the product of the two matrices, with the matrix of the first transformation on the right. The identity transformation is represented by the unit matrix $\begin{pmatrix} 1 & 0 \\ 0 & 1 \end{pmatrix}$, and the inverse of a transformation is represented by the reciprocal of the corresponding matrix. The matrices are the elements of the group $SL(2, \mathbb{R})$; hence the isometries of the model are represented by the quotient group $SL(2, \mathbb{R})/(I, -I)$, where I represents the unit matrix; see Note below. We shall see below that these matrices thus represent *all* the direct isometries of the hyperbolic plane. A particular reverse isometry is the reflection $z \to -\bar{z}$; hence the reverse isometries are represented by transformations of the form $z \to z' = (a\bar{z} + b)/(c\bar{z} + d)$, where $ad - bc = -1$.

Note: It is recalled from linear algebra that an $n \times n$ matrix with real entries represents a linear transformation in an n-dimensional real space $\mathbb{R}^n$, and the nonzero matrices represent the *general linear group* $GL(n, \mathbb{R})$. The matrices with determinant 1 represent the *special linear group* $SL(n, \mathbb{R})$.

Let r, θ be polar coordinates in the hyperbolic plane, so chosen that the origin is mapped onto the point $z = i$ of the model and the "horizontal" line (where $\theta = 0$ or π) is mapped onto the semicircle $|z| = 1$ of the model. Let R_α and T_X denote a rotation in the plane about the origin by angle α and a translation by distance X along the "horizontal" line. It is easy to show (see Exercises in Section 7.8) that those isometries are represented in the half-plane model by mappings (8.1-1) with matrices as follows:

$$\text{Rotation} \quad R_\alpha : \quad \begin{pmatrix} \cos(\alpha/2) & \sin(\alpha/2) \\ -\sin(\alpha/2) & \cos(\alpha/2) \end{pmatrix} \tag{8.1-2}$$

$$\text{Translation} \quad T_X : \quad \begin{pmatrix} \cosh(X/2) & \sinh(X/2) \\ \sinh(X/2) & \cosh(X/2) \end{pmatrix} \tag{8.1-3}$$

As shown by Theorem 3.15 of Section 3.6, the general direct isometry can be written in the form $\sigma = R_{\alpha'} T_X R_\alpha$, and we write the corresponding matrix P as the product of the corresponding matrices (8.1-2,3). With $P = \begin{pmatrix} a & b \\ c & d \end{pmatrix}$, we find

$$a = \cosh \frac{X}{2} \cos \gamma + \sinh \frac{X}{2} \sin \delta,$$

$$b = \cosh \frac{X}{2} \sin \gamma + \sinh \frac{X}{2} \cos \delta,$$

$$c = - \cosh \frac{X}{2} \sin \gamma + \sinh \frac{X}{2} \cos \delta,$$

$$d = \cosh \frac{X}{2} \cos \gamma - \sinh \frac{X}{2} \sin \delta,$$

$$(8.1\text{-}4)$$

where

$$\gamma = \frac{\alpha' + \alpha}{2}, \qquad \delta = \frac{\alpha' - \alpha}{2}.$$

To solve these equations for $\cosh \frac{X}{2}$, $\sinh \frac{X}{2}$, γ, and δ, we find, first,

$$\cosh^2 \frac{X}{2} = \left(\frac{a+d}{2} \right)^2 + \left(\frac{b-c}{2} \right)^2,$$

$$\sinh^2 \frac{X}{2} = \left(\frac{a-d}{2} \right)^2 + \left(\frac{b+c}{2} \right)^2,$$

$$\tan \gamma = \frac{b-c}{a+d}$$

$$\tan \delta = \frac{a-d}{b+c}.$$

$$(8.1\text{-}5)$$

These equations give $\cosh^2(X/2) - \sinh^2(X/2) = ad - bc$, but $\cosh^2 - \sinh^2$ is always 1; hence we must require that $ad - bc = 1$, and then we can solve for $X, \gamma, \delta, \alpha, \alpha'$.

Conclusion: Every direct isometry of the hyperbolic plane is represented by a real 2×2 matrix of determinant 1, and, conversely, every such matrix represents an isometry of the hyperbolic plane. (As noted earlier, a matrix P and its negative $-P$ represent the same isometry.)

Recall that the symmetric matrix T_X represents a translation along the line $\theta = 0$. If $\alpha' = -\alpha$, then P given as above represents a translation along a line through the origin at angle $\theta = \alpha$, and P is also a symmetric matrix.

In Chapter 10, as a by-product of the study of Lorentz transformations, it will be shown that the isometries of the three-dimensional hyperbolic space $\mathbb{H}^3$ are similarly represented by *complex* 2×2 matrices of determinant 1.

8.2 Points, Lines, and Curves Invariant Under an Isometry

A point is *invariant* under a mapping if it is mapped onto itself. A line is *invariant* if each of its points is mapped onto a (generally different) point of the same line, so that the line as a whole is unaltered by the mapping. (Note that if the mapping is an isometry, hence is a one-to-one mapping of the whole plane, the image is the entire line without gaps.) The same applies to invariance of a curve.

In the Euclidean plane, certain points, lines, and circles are invariant under certain isometries. In the hyperbolic plane, there are certain other curves that are also invariant under certain isometries, namely, equidistants and horocycles.

The existence of an invariant point or line or curve generally determines the corresponding isometry, up to a single parameter. For example, if a point P_0 is invariant under an isometry, then, since the isometry maps any figure onto a congruent figure, lines through P_0 are mapped onto lines through P_0, if two lines through P_0 make an angle θ, their images make the same angle, and the distances $|PP_0|$ from P_0 to any other points P are invariant, so that the isometry is either a rotation through some angle about P_0 or a reflection in some line through P_0. The parameter referred to above is the angle of rotation in the first case and the coordinate θ of the line of reflection in the second. From now on, we shall assume that all isometries we are concerned with are direct, so there are no reflections.

Exercises

1. Show that if a direct isometry leaves a line invariant, then it is a translation through some distance X along that line.

2. Show that if a direct isometry leaves two finite points fixed, it is the identity. Show that if it leaves two distinct ideal points at infinity fixed, it is a translation through some distance X along the line that joins the two ideal points. What happens, if there is one finite fixed point and one ideal fixed point?

For the further analysis of invariant figures, it is now convenient to make use of the Poincaré half-plane model, which is now justified, because we have proved that all models are isomorphic with each other and with the abstract hyperbolic plane.

As noted in the preceding section, the isometries of the hyperbolic plane are represented by fractional linear transformations $z \to z' = (az + b)/(cz + d)$, where a, b, c, d are real constants such that $ad - bc = 1$. A fixed point in the hyperbolic plane gives a fixed point of the model, which is a point z such that $z = (az + b)/(cz + d)$, that is, it is a root of the quadratic equation

$$cz^2 + (d - a)z - b = 0. \qquad (8.2\text{-}1)$$

Since $ad - bc = 1$, the discriminant is equal to $(a + d)^2 - 4$; hence the fixed points are given, in case $c \neq 0$, by

$$z = \frac{a - d \pm \sqrt{(a + d)^2 - 4}}{2c}. \tag{8.2-2}$$

We divide the discussion into two cases — case I, in which $c \neq 0$, with subcases Ia, Ib, Ic, and case II, in which $c = 0$, with subcases IIa, IIb, and IIc.

Ia: $(d + a)^2 > 4$. Equation (8.2-1) has two real roots, which represent fixed ideal points at infinity. Hence there is an invariant line, and the isometry is a translation along that line.

Ib: $(d + a)^2 = 4$. There is one fixed ideal point and no fixed finite points. As we shall see, the isometry is an ideal rotation.

Ic: $(d + a)^2 < 4$. There is just one fixed point in the upper half-plane, and the isometry is a rotation.

For $c = 0$, the mapping is linear $z \rightarrow z' = (az + b)/d$. As $z \rightarrow \infty$, z' also $\rightarrow \infty$; hence the ideal point at $\theta = (\pi/2)$ is fixed. The quadratic equation (8.2-1) is a linear equation, and we have to consider three cases:

IIa: $d - a \neq 0$. There is one real fixed point in the model; hence there are two fixed ideal points (the other is at $\theta = (\pi/2)$), and the isometry is a translation.

IIb: $d - a = 0$, $b \neq 0$. Then, since c is also zero, (8.2-1) has no solution. There is no finite fixed point in the model; hence there is just one fixed ideal point (the one at $\theta = (\pi/2)$). Again, the mapping is an ideal rotation; see discussion below.

IIc: $d - a = 0$, $b = 0$. The mapping in the model is the identity; hence the isometry is the identity, in which all points are fixed.

We now analyze the ideal rotations in more detail. They are the cases in which there is just one fixed ideal point in the hyperbolic plane, and no finite fixed points; they were discussed in Section 3.6 and 3.8. Since the geometry of the plane is symmetric with respect to rotations about the origin, we can choose the location at infinity of the fixed ideal point to suit our convenience, and it is convenient to choose it at $\theta = (\pi/2)$. That gives Case IIb above, in which the fractional linear transformation reduces to the horizontal translation $z \mapsto z + b$ in the model. We recall that the lines in the plane are represented, in the model, by semicircles centered on the real axis, together with vertical half-lines; the latter represent an asymptotic pencil of lines all going to the ideal point at $\theta = (\pi/2)$. In the mapping in the model, each of the lines of that pencil is mapped onto another such at a distance b horizontally; hence the isometry is an ideal rotation, as described in Section 3.6. In the model, all horizontal lines, $y = $ a positive constant, are invariant, and they represent horocycles in the plane. Those horocycles are perpendicular to the lines of the pencil, because they are perpendicular to the vertical lines in the model, and the model is conformal. As said in

Section 3.9, each horocycle is a curve in the hyperbolic plane that "slides along itself" under the ideal rotation.

We recall that, given a line ℓ, a curve C is an *equidistant* to ℓ if the distance d from a point Q on C to ℓ (as measured by dropping a perpendicular) is the same for all points Q on C. Clearly, the curve C is invariant under translations along ℓ, because, under that isometry, the segment PQ is moved to another segment $P'Q'$ also perpendicular to ℓ and of the same length d, as in Fig. 8.2a.

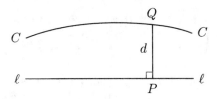

Fig. 8.2a.

Further Exercises

3. Show that each of the following is a subgroup of the isometry group:
   a. The set of all rotations about a given point P_0.
   b. The set of all translations along a given line ℓ.
   c. The set of all ideal rotations involving a given ideal point.
4. Describe the composition of two ideal rotations involving different ideal points at infinity.
5. Show that the ideal rotation $T_y R_{2\beta} T_y$ discussed in Exercise 3 of Section 3.8 is represented in the half-plane model by a matrix of the form $\left(\begin{smallmatrix} 1 & b \\ 0 & 1 \end{smallmatrix}\right)$. *Hint:* $\beta = (\pi/2) - \alpha$ and $\alpha = \Pi(y)$.
6. Show that under any isometry (rotation, translation, ideal rotation, or reflection), every point of the plane is on some line or curve that is invariant (except for a center of rotation, unless you are willing to regard that point as a circle of radius zero).
7. Show how to modify the arguments of this section so as to include reverse isometries (i.e., ones that include a reflection).

8.3 Simplicity of the Hyperbolic Direct Isometry Group

We wish to show that the Euclidean and hyperbolic planes have fundamentally different direct isometry groups, which we denote by Γ_E and Γ_h, respectively, in this section.

The general direct isometry of the Euclidean plane is given by

$$\begin{pmatrix} x \\ y \end{pmatrix} \rightarrow \begin{pmatrix} x' \\ y' \end{pmatrix} = \begin{pmatrix} \cos\theta & \sin\theta \\ -\sin\theta & \cos\theta \end{pmatrix} \begin{pmatrix} x \\ y \end{pmatrix} + \begin{pmatrix} a \\ b \end{pmatrix}, \qquad (8.3\text{-}1)$$

where x and y are Cartesian coordinates (which exist in the Euclidean but not the hyperbolic plane); it represents a rotation (counter-clockwise) through an angle θ followed by a translation by distance a in the x-direction and a distance b in the y-direction. (The same can also be obtained by a suitable translation *followed* by a rotation.) It is convenient to represent the above isometry by a 3×3 matrix, by writing

$$\begin{pmatrix} x \\ y \\ 1 \end{pmatrix} \rightarrow \begin{pmatrix} x' \\ y' \\ 1 \end{pmatrix} = \begin{pmatrix} \cos\theta & \sin\theta & a \\ -\sin\theta & \cos\theta & b \\ 0 & 0 & 1 \end{pmatrix} \begin{pmatrix} x \\ y \\ 1 \end{pmatrix}. \qquad (8.3\text{-}2)$$

It is clear that the composition of two such isometries is represented by the matrix product of the two corresponding 3×3 matrices, the matrix of the first isometry appearing on the right. (See Exercise 1 below.)

In this case (the Euclidean case), the pure translations are represented by matrices of the form

$$T = T_{a,b} = \begin{pmatrix} 1 & 0 & a \\ 0 & 1 & b \\ 0 & 0 & 1 \end{pmatrix}. \qquad (8.3\text{-}3)$$

The composition of two pure translations (in either order) is another pure translation obtained by adding the translation vectors: $a'' = a + a'$, $b'' = b + b'$. (See Exercise 2 below.) Hence the translations constitute a subgroup G_0 (which is commutative or Abelian) of the Euclidean group Γ_E.

Furthermore, the subgroup G_0 is a normal subgroup of Γ_E. Recall that a subgroup G_0 of a group G is called *normal* if it is invariant under all conjugations of the group, that is, if for all P in the subgroup G_0 and all Q in the whole group G, the element $Q^{-1}PQ$ is always in the subgroup G_0. It is easily verified that if Q is any 3×3 matrix of the form (8.3-2), then $Q^{-1}T_{a,b}Q$ can be written as $T_{a',b'}$, hence the subgroup of pure translations is normal. (See Exercise 3.)

In the hyperbolic case, the pure translations do not constitute a subgroup of Γ_h; furthermore, Γ_h does not contain any nontrivial proper normal subgroup. (A subgroup is *nontrivial* if it is not merely the identity subgroup $\{e\}$ consisting of the identity element e of the group alone, and it is *proper* if it is not all of the group G.) If a group G does not contain a nontrivial proper normal subgroup, it is called *simple*. (A simple group may contain nontrivial proper subgroups, but they are not normal; for example, the set of all translations along a given line ℓ is a subgroup of Γ_h, but it is not normal.) The main theorem of this section is this:

Theorem 1: Γ_h is a simple group.

The remainder of this section is devoted to the proof of this theorem. According to the preceding two sections, an element of Γ_h can be represented (via the Poincaré half-plane model) by a pair $(P, -P)$, where P is a 2×2 real matrix of determinant 1, i.e., an element of $SL(2, \mathbb{R})$.

We recall from linear algebra that a matrix M with distinct eigenvalues can be diagonalized by a conjugation: $Q^{-1}MQ = D$ (diagonal). The following is a consequence.

Lemma 1: If A and B are any real 2×2 matrices of determinant 1 (i.e., elements of $SL(2, \mathbb{R})$ with the same eigenvalues λ_1 and λ_2, where $\lambda_1 \neq \lambda_2$, then there is a real matrix Q with determinant 1 such that

$$Q^{-1}AQ = B, \tag{8.3-4}$$

(*Note:* It is not assumed that λ_1 and λ_2 are real.)

Proof: There are matrices P, R (not necessarily real) such that

$$P^{-1}AP = \begin{pmatrix} \lambda_1 & 0 \\ 0 & \lambda_2 \end{pmatrix}, \qquad R^{-1}BR = \begin{pmatrix} \lambda_1 & 0 \\ 0 & \lambda_2 \end{pmatrix}; \tag{8.3-5}$$

hence,

$$Q^{-1}AQ = B, \quad \text{where} \quad Q = PR^{-1}. \tag{8.3-6}$$

The matrix Q is not unique; it can be multiplied through by any nonzero constant. Since Q satisfies a linear equation $AQ - QB = 0$, we see from the Gauss elimination procedure that if the multiplicative constant is so chosen as to make one matrix entry of Q real, then those entries are all real. Furthermore, since $\det A = \det B = 1$, we see from the above equation that the multiplicative constant can be such that $\det Q = \pm 1$. Last, if $\det Q = -1$, we interchange λ_1 and λ_2 in *one* of equation (8.3-5). That has the effect of multiplying Q by $\begin{pmatrix} 0 & 1 \\ 1 & 0 \end{pmatrix}$, hence changing the determinant of Q to $+1$ (because $\det\begin{pmatrix} 0 & 1 \\ 1 & 0 \end{pmatrix} = -1$). $\qquad\square$

Lemma 2: If A is any real 2×2 matrix with $\det A = 1$, its eigenvalues are

$$\frac{\operatorname{tr} A \pm \sqrt{(\operatorname{tr} A)^2 - 4}}{2}, \tag{8.3-7}$$

where tr denotes trace; $\operatorname{tr}\begin{pmatrix} a & b \\ c & d \end{pmatrix} = a + d$.

Proof: Elementary.

Hence it is convenient to distinguish three cases: $|\operatorname{tr} A| < 2$, $|\operatorname{tr} A| = 2$, and $|\operatorname{tr} A| > 2$. Recall that each element of Γ_h is a pair $(P, -P)$. In order to prove that Γ_h is simple (Theorem 2 above), we have to prove that if a normal subgroup G_0 contains even one element $(P, -P)$ not equal to the identity $(I, -I)$, then it contains all elements of Γ_h. We do that in three parts:

Proposition 1: If G_0 contains an element $(A, -A)$ with $0 \leq \operatorname{tr} A < 2$ [hence $-2 < \operatorname{tr}(-A) \leq 0$], then it contains all elements $(P, -P)$ of Γ_h.

Proposition 2: If G_0 contains an element $(A, -A)$ with $\operatorname{tr} A > 2$ [hence $\operatorname{tr}(-A) < -2$], then it contains all elements of Γ_h.

Proposition 3: If G_0 contain an element $(A, -A)$ with $\operatorname{tr} A = 2$ [hence $\operatorname{tr}(-A) = -2$] and $A \neq \pm I$, then G_0 is all of Γ_h.

Proof of Proposition 1: Assume that G_0 contains an element $(A, -A)$, where $\operatorname{tr} A < 2$. According to Lemma 2, if two matrices have the same trace, then they have the same eigenvalues, and we see from Lemma 1 that G_0 contains all elements $(P, -P)$ with $\operatorname{tr} P = \operatorname{tr} A$. In particular, G_0 contains the element $(B, -B)$ where

$$B = \begin{pmatrix} \cos\alpha & \sin\alpha \\ -\sin\alpha & \cos\alpha \end{pmatrix}, \quad \text{where} \quad \cos\alpha = \frac{1}{2}\operatorname{tr} A.$$

Then, G_0 contains all elements with matrices of the form $Q^{-1}BQ$. We choose $Q = \begin{pmatrix} b & 0 \\ 0 & 1/b \end{pmatrix}$, so that G_0 contains an element with the matrix

$$C_1 = \begin{pmatrix} \cos\alpha & b^{-2}\sin\alpha \\ -b^2\sin\alpha & \cos\alpha \end{pmatrix}. \tag{8.3-8}$$

Then, G_0 contains the element obtained from this one by replacing α by $-\alpha$, because that is equivalent to replacing B by B^{-1}, and any subgroup contains the inverses of all its elements. We can also replace b by $1/b$ (because b was arbitrary). If we call the result C_2, then the subgroup contains the element with matrix

$$C_1 C_2 = \begin{pmatrix} \cos^2\alpha + b^4\sin^2\alpha & -\sin\alpha\cos\alpha(b^2 - b^{-2}) \\ -\sin\alpha\cos\alpha(b^2 - b^{-2}) & \cos^2\alpha + b^{-4}\sin^2\alpha \end{pmatrix}.$$

The trace of this matrix is $2\cos^2\alpha + (b^4 + b^{-4})\sin^2\alpha$, which can be made equal to any number $2\cos^2\alpha + (b^4 + b^{-4})\sin^2\alpha$, which can be made equal to any number > 2 by suitable choice of b. Hence a partial result is that G_0 contains all $(P, -P)$ with $\operatorname{tr} P > 2$. In particular, it contains all translations in the plane (by any distance along any line). To see that, recall from Section 8.1 that a translation by a distance X to the right along the θ-axis (of the polar coordinate system) is represented by the symmetric matrix

$$T_X = \begin{pmatrix} \cosh(X/2) & \sinh(X/2) \\ \sinh(X/2) & \cosh(X/2) \end{pmatrix}. \tag{8.3-9}$$

A translation by distance X along any line anywhere in the plane is given by a matrix of the form $Q^{-1}T_X Q$. Since eigenvalues are invariant under conjugation, this matrix has the same trace as T_X, namely, $2\cosh(X/2)$, hence is contained in the subgroup. $\qquad \square$

We are now in a position to show that G_0 contains all elements $(P, -P)$ with $\operatorname{tr} P$ in the interval $[0, 2)$. We recall from Section 3.7 that if ABC is any triangle, and if we perform three translations in succession, namely, by a distance $|AB|$ along $\overrightarrow{AB}$, a distance $|BC|$ along $\overrightarrow{BC}$, and a distance $|CA|$ along $\overrightarrow{CA}$, the resulting isometry returns vertex A to its original position and is a rotation about that vertex through an angle α equal to the angular defect of the triangle ABC. The matrix that represents that rotation (taking A as the origin of polar coordinates) in the half-plane model is

$$\begin{pmatrix} \cos(\alpha/2) & \sin(\alpha/2) \\ -\sin(\alpha/2) & \cos(\alpha/2) \end{pmatrix}.$$

Since α can be any angle between 0 and π, the trace of this matrix can have any value between 0 and 2. (To take care of the limiting case $\alpha = \pi$ (trace $= 0$), consider translations around the sides of a quadrilateral, for then the defect can be anything between 0 and 2π.

To complete the proof of Proposition 1, we must show how to include matrices of trace 2. An example is $\begin{pmatrix} 1 & 1 \\ 0 & 1 \end{pmatrix}$, which represents an ideal rotation or "rotation about the ideal point at infinity in the direction $\theta = (\pi/2)$." We wish to show that $\begin{pmatrix} 1 & 1 \\ 0 & 1 \end{pmatrix}$ is in the subgroup. Namely,

$$\begin{pmatrix} 1 & 1 \\ 0 & 1 \end{pmatrix} = \begin{pmatrix} 0 & -1 \\ 1 & 0 \end{pmatrix} \begin{pmatrix} 0 & 1 \\ -1 & -1 \end{pmatrix};$$

this is the product of two matrices neither of which has trace 2; hence they are in the subgroup by what has already been proved; hence $\begin{pmatrix} 1 & 1 \\ 0 & 1 \end{pmatrix}$ is in the subgroup. Hence so are also its conjugates, which represent rotations about ideal points in other directions at infinity. According to Lemma 2, any matrix with trace 2 (and determinant 1) has eigenvalues both equal to 1. By the general theory, it is then conjugate either to the unit matrix I (in which case it is equal to I) or to the Jordan matrix $\begin{pmatrix} 1 & 1 \\ 0 & 1 \end{pmatrix}$. Hence, G_0 contains all matrices of trace 2. This completes the proof of Proposition 1.

Proof of Proposition 2: By hypothesis, G_0 contains a matrix with trace $>$ 2, that is, a matrix conjugate to T_X given by (8.3-9) for some value of X. We consider an equilateral triangle of side length X and angle α, as in Fig. 8.3. The angle α is related to X by the equation $\tanh(X/2) = \tanh X \cos \alpha$, but in any case, by the argument in the proof of Proposition 1, we see that G_0 contains a rotation through the angular defect $(ABC) = \pi - 3\alpha$. The trace of the corresponding matrix is less than 2; hence, from now on, we repeat the proof of Proposition 1.   □

Proof of Proposition 3: By hypothesis, G_0 contains an element $(P, -P)$, where P has trace 2, and $P \neq I$. By the argument at the end of the proof of Proposition 1, G_0 therefore contains all matrices of trace $= 2$, in particular the matrices $\begin{pmatrix} 1 & k \\ 0 & 1 \end{pmatrix}$ and $\begin{pmatrix} 1 & 0 \\ k & 1 \end{pmatrix}$. Hence, it contains the product

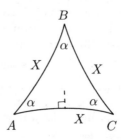

Fig. 8.3.

$$\begin{pmatrix} 1 & k \\ 0 & 1 \end{pmatrix}\begin{pmatrix} 1 & 0 \\ k & 1 \end{pmatrix} = \begin{pmatrix} 1+k^2 & k \\ k & 1 \end{pmatrix},$$

which has trace > 2, so that from here on, we repeat the proof of Proposition 2.   □

The main theorem (Theorem 1 above) follows from these three propositions.

Exercises

These exercises are all for the Euclidean case $\mathbb{E}^2$.

1. Show that if the transformation (8.3-2) is followed by a similar transformation from x', y' to x'', y'', the composition of the two mappings is again of the same form, and its matrix is the product of the two matrices, the matrix of the first transformation appearing on the right.

2. Show that the product of two translation matrices of the form (8.3-3) is a translation matrix, representing a translation obtained by vector addition of the vectors $\binom{a}{b}$ and $\binom{a'}{b'}$.

3. Show that for any $T_{a,b}$ of the form (8.3-3) and any Q of the form (8.3-2), $Q^{-1}T_{a,b}Q$ is again a translation matrix $T_{a',b'}$, for some a', b'.

8.4 The Group $SL(2, \mathbb{Z})$ and the Corresponding Tiling

The group $SL(2, \mathbb{Z})$ consists of 2×2 matrices M of the form $\begin{pmatrix} a & b \\ c & d \end{pmatrix}$ whose entries a, b, c, d are integers such that $ad - bc = 1$. The product of two such matrices is again of that form, and the inverse of $\begin{pmatrix} a & b \\ -c & d \end{pmatrix}$ is the matrix $\begin{pmatrix} d & -b \\ c & a \end{pmatrix}$; hence those matrices M form a subgroup of the group $SL(2, \mathbb{R})$, and the corresponding fractional linear transformations $z \to z' = (az + b)/(cz + d)$ of the upper half-plane represent a subgroup of the group of isometries of the hyperbolic plane.

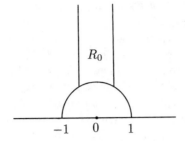

Fig. 8.4a.

Let R_0 denote the region of the upper half-plane bounded by the segment of the unit semicircle extending from $x = -(1/2)$ to $x = +(1/2)$ and the vertical half-lines extending upward to infinity from the endpoints of that segment, as in Fig. 8.4a. The region R_0 represents a triangle in the hyperbolic plane with $60°$ angles at two vertices separated by the distance $\ln 3$ and its third vertex an ideal point at infinity, where the angle is $0°$ (the two sides are asymptotic at that ideal point). Under the mappings $z \to z' = z+b$ corresponding to matrices $\left(\begin{smallmatrix} 1 & b \\ 0 & 1 \end{smallmatrix}\right)$ the region R_0 is shifted to the right or left by the integer distance b, to give a sequence $\dots, R_{-1}, R_0, R_1, \dots$ of adjacent regions, as shown in Fig. 8.4b. We let R_0' be the region obtained by inversion of R_0 in the unit semicircle, as shown; it represents the triangle in the hyperbolic plane obtained by reflection of the triangle represented by R_0 in the line of its base, which is another of the triangles in the "reflection triangulation" described in Section 3.11. Similarly, let R_i' be the region underneath R_i in the same way, as in the drawing. Then, the portion of the upper half-plane not yet covered consists of a succession of adjacent regions $\dots, \Omega_{-1}, \Omega_0, \Omega_1, \dots$, each of which is a convex region bounded below by a portion of the real axis and two curves which meet above at an angle of $120°$.

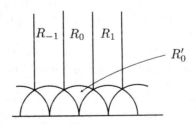

Fig. 8.4b.

In the region $\Omega_0 = ABC$ shown in Fig. 8.4c, we consider points D, E obtained by going down the circular arcs AC and BC to points D, E at a distance from C that corresponds to a distance $\ln 3$ in the hyperbolic plane, and then we draw the circular arcs from D and E meeting the midpoint F of the base AB at right angles. The new regions so obtained CDF and CEF represent further triangles of the reflection triangulation of the hyperbolic plane. We do the same thing to the other regions Ω_i. After that, the portion of the upper half-plane not yet covered consists again of a succession of adjacent regions $\ldots, \Omega'_{-1}, \Omega'_0, \Omega'_1, \ldots$, each of which has the same character as the Ω_i; namely, it is a convex region bounded by a segment of the real axis and two curves meeting above at $120°$. We treat these regions in the same way as the Ω_i, and so on. In the limit the upper half-plane is covered by nonoverlapping regions which represent the reflection triangulation of the hyperbolic plane, a portion of which is shown schematically in Fig. 8.4d.

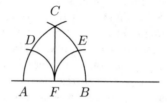

Fig. 8.4c.

This tiling differs from the ones considered in Section 3.11 in that each triangle has a thin finger reaching out to infinity among the other triangles. We now investigate the locations of the points at infinity at which "vertices" of the "triangles" are to be found. As usual, we assume that r, θ are polar coordinates in $\mathbb{H}^2$ with origin at the point represented by the point $z = i$ in the model and with the axis where $\theta = 0$ represented by the unit semicircle that passes through the point $z = i$. Then, one of the ideal points involved is the one at infinity in the model, which corresponds to $\theta = (\pi/2)$. The other ideal points can be found by applying the fractional linear transformations. If we let $z \to \infty$, then in the limit, we have $z' = (a/c)$, which is a rational point on the real axis. Furthermore, any rational number on the real axis can be written as (a/c) in lowest terms, and then, since a and c are relatively prime, there are integers c and d such that $ad - bc = 1$, and we see that *all* rational points on the real axis represent ideal points where sides of the triangles meet. To find the corresponding angles θ, we use equation (7.8-9), so that

$$\frac{a}{c} = \frac{1 + \sin\theta}{\cos\theta}. \qquad (8.4\text{-}1)$$

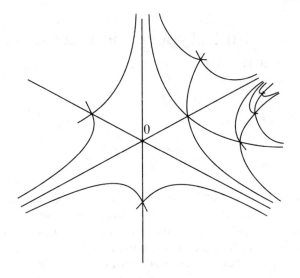

Fig. 8.4d.

If we square both sides and replace $\cos^2\theta$ by $1-\sin^2\theta$, we obtain a quadratic equation for $\sin\theta$. The discriminant of that equation is $4a^4$; hence $\sin\theta$ is rational, and it then turns out that $\cos\theta$ is also rational.

This group and tiling play a role in the theory of the elliptic modular function $J(z)$, which may be regarded either as a meromorphic function (a function analytic everywhere except for isolated poles) in the hyperbolic plane or as such a function in the upper half-plane of the model. In each triangle of the tiling, $J(z)$ takes on each complex value exactly once (for this purpose, some of the boundary points of the triangles have to be included) and has a single pole of the first order. The function is invariant under each of the reflections that generate the tiling. See L. R. Ford, *Automorphic Functions*, Chelsea, New York, 1951, or Whittaker and Watson, *A Course of Modern Analysis*, 4th ed., Cambridge University Press, London and New York, 1935.

Exercises

1. From the formula for the angle of parallelism, verify that the bases of the triangles referred to have length $\ln 3$.
2. From (8.4-1) find expressions for $\sin\theta$ and $\cos\theta$. Is it true that $\sin\theta$ and $\cos\theta$ can be any rational numbers the sum of whose squares is 1?
3. Are *all* fractional linear transformations corresponding to matrices in $SL(2,\mathbb{Z})$ involved? (*Recall*: A direct isometry comes from two reflections.)

Chapter 9

Differential and Hyperbolic Geometry in More Dimensions

The differential geometry of Chapter 5 is extended to spaces with more than two dimensions. General manifolds are defined, and a space is determined by a manifold equipped with a line element or metric tensor. In particular, the line element for three-dimensional hyperbolic space is given in terms of the spherical coordinates, and a second proof that the geometry of the horosphere is Euclidean is given. Various models of three-dimensional hyperbolic space are described. Lastly, spaces with indefinite metric, in which the metric tensor is not required to be a positive definite matrix, are discussed. A particular case of this is the Minkowski geometry used in the theory of special relativity. The Minkowski geometry appears in the next chapter in connection with the relation between hyperbolic isometries and Lorentz transformations of relativity theory.

9.1 Manifolds

It was noted in Section 5.1 that to describe a complete closed surface like a sphere or the surface of a torus it is generally necessary to define two or more coordinate charts and apply certain smoothness conditions in the overlap of any two of them. These conditions are stated in terms of coordinate changes being of class C^k, which means that the functions have continuous partial derivatives of all orders less than or equal to k.

We now define an *n-dimensional manifold* as follows: Let $\mathcal{M}$ be a metric space, simply so that we can talk about open sets and continuity. If V_1 is an open subset of $\mathcal{M}$, and if in V_1 coordinates $u^1, \ldots, u^n$ are defined in such a way as to give a homeomorphism of V_1 to a convex region Ω of an n-dimensional parameter space ($u^1, \ldots, u^n$ are regarded as Cartesian coordinates in the parameter space), we say that we have a *coordinate chart* in $\mathcal{M}$. Suppose V_2 is another coordinate chart with coordinates $v^1, \ldots, v^n$. Then in the overlap of the charts (i.e., in the intersection of the subsets V_1 and V_2 of $\mathcal{M}$), the parameters $u^1, \ldots, u^n$ are single-valued continuous functions of $v^1, \ldots, v^n$, and conversely, because both sets of coordinates determine homeomorphisms. Thus two charts are automatically compatible in the topological sense. Finally, if $\mathcal{M}$ is covered by a collection of charts, then we have an n-dimensional manifold.

If such a manifold satisfies the further requirement that in the overlap of any two charts, the coordinates of the first chart (say $u^1, \ldots, u^n$) are C^k functions of the coordinates of the second chart (say $v^1, \ldots, v^n$), and conversely, then we say that we have a manifold *of class* C^k. Here we have $1 \leq k \leq \infty$.

Another method of describing a manifold is the method of *identification of points* on the boundary of a region Ω. For example, let Ω be the unit square in the u, v-plane: $0 \leq u \leq 1$, $0 \leq v \leq 1$. We then identify the points on the vertical sides by decreeing that for each v, the points $(0, v)$ and $(1, v)$ are the same point of the surface being described. To visualize this, think of Ω as a flat square of paper, which we roll up to make into a tube by gluing the sides together. Then, we identify the points on top and bottom by decreeing that for each u the points $(u, 0)$ and $(u, 1)$ are the same point. To visualize that, we suppose the paper to be sufficiently flexible that we can bend the tube around and join its ends to make a torus, as indicated schematically in Fig. 9.1a. (To enforce smoothness, as for a manifold of class C^2, we can suppose the edges of the region to overlap slightly before gluing; that is, we identify not merely points on the boundary of the square, but in a suitable manner the points in thin strips at the boundary.)

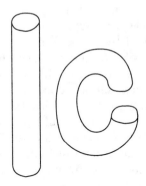

Fig. 9.1a.

An interesting variation of the above construction of the torus is this: In the second step, we identify, for each u, the points $(u, 0)$ and $(1 - u, 1)$. That gives an interesting surface called a *Klein bottle*, which can be visualized as follows: We start with a bottle with a long slender neck and a circular opening in the bottom, as on the left in Fig. 9.1b.

We then bend the neck downward and make it penetrate the side of the bottle and join it inside to the opening in the bottom, thus creating

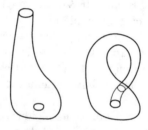

Fig. 9.1b.

a surface without boundary, but having *only one side*: If we start at a point P on the outside of the neck, we can follow it down to a point inside the bottle and from there back up inside the neck to the point P on the neck, but we are now on the inside of the surface. Although the Klein bottle cannot be embedded in 3-space $\mathbb{E}^3$ without self-intersection (where the neck penetrates the side of the bottle), it can be embedded without self-intersection in $\mathbb{E}^4$. When it is embedded in $\mathbb{E}^3$ as described above, we have three numbers x, y, z at each point of the surface, in such a way that they vary continuously on the surface. We then attach a fourth number w to each point in such a way that w also varies smoothly on the surface, but takes on different values on the neck and side at the points of intersection. Then, in $\mathbb{E}^4$ with coordinates x, y, z, w, there is no self-intersection.

As another example of the method of identification, we consider the manifold of the group G of rotations of a rigid body about a fixed point of $\mathbb{E}^3$: Each rotation can be described by a vector from the origin of a parameter space in the direction of the axis of rotation and of magnitude equal to the angle of rotation, hence of length in the interval $[0, \pi]$. The tips of these vectors fill up a spherical ball Ω of radius π, but diametrically opposite points on the surface of this ball have to be identified, because a rotation through angle π gives the same effect as rotation through angle π in the opposite direction. (We are not concerned with the process of rotation, but only with the final orientation of the solid body; each element of the group represents such a final orientation.) We get in this way a three-dimensional manifold, the *manifold* of the rotation group; it can also be embedded in $\mathbb{E}^9$ without self-intersection (see Exercise 1 below).

Exercises

1. Show how the manifold of the rotation group can be embedded in $\mathbb{E}^9$ without self-intersection.
2. Suppose that in the square referred to for the torus and the Klein bottle, we do not identify points of the vertical sides, but we identify the top and bottom by identifying each point $(u, 0)$ with $(1 - u, 1)$. What well-known surface results?

9.2 The Line Element, Geodesics, Volume

If a metric tensor (an $n \times n$ symmetric, positive definite matrix) is defined in each chart of a manifold of class C^2 in such a way that the transformation law (5.2-8) holds in the intersection of any two charts (with the number of coordinates increased from 2 to n), we say that we have a *Riemannian manifold*. Two-dimensional Riemannian manifolds are the abstract geometric surfaces of interest in connection with the hyperbolic plane $\mathbb{H}^2$. In most models of $\mathbb{H}^2$ the entire plane is covered by one coordinate chart (the exception being the pseudosphere, which represents only a portion of $\mathbb{H}^2$), but we often consider different coordinate systems, so the transformation law (5.2-8) is still of importance.

The line element ds is given by the obvious generalization of (5.2-3):

$$ds^2 = \sum_{j,k=1}^{n} g_{jk} du^j du^k, \qquad (9.2\text{-}1)$$

The equation of a geodesic is the same as (5.6-6), but with j, k, m all going from 1 to n; the Christoffel three-index symbols are still given by (5.6-4) and (5.6-5).

The volume of a region is given by the obvious generalization of (5.8-1) for area. In the three-dimensional case ($n = 3$), we have

$$\text{Volume}\,(\mathcal{R}) = \int \int \int_{\Omega} \sqrt{\det G}\; du^1 du^2 du^3. \qquad (9.2\text{-}2)$$

9.3 The Line Element in Hyperbolic 3-Space

In any plane in $\mathbb{H}^3$ the hyperbolic line element holds, which we found to be

$$ds^2 = dr^2 + (\sinh r)^2 d\theta^2, \qquad (9.3\text{-}1)$$

in polar coordinates r, θ. Spherical coordinates ρ, ϕ, θ were defined in $\mathbb{H}^3$ in Section 4.4. In a meridinal half-plane $\theta = \text{const}$, increments $d\rho$ and $d\phi$ yield a displacement ds_1 given, according to the above equation, by

$$ds_1^2 = d\rho^2 + (\sinh \rho)^2 d\phi^2. \qquad (9.3\text{-}2)$$

The displacement that results from an increment $d\theta$ at constant ρ, ϕ is perpendicular to the meridional plane, that is, perpendicular to the plane of Fig. 9.3, and is in a plane perpendicular to the polar axis, and the distance d from the axis in that plane is given, according to Chapter 6, by $\sinh d = \sinh \rho \sin \phi$; hence the corresponding displacement is given by

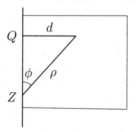

Fig. 9.3.

$$ds_2^2 = (\sinh d)^2 d\theta^2 = (\sinh \rho \sin \phi)^2 d\theta^2. \qquad (9.3\text{-}3)$$

Since the displacements ds_1 and ds_2 are orthogonal, and the Pythagorean theorem holds in the limit of small figures, the line element in $\mathbb{H}^3$ is given by the sum of the two expressions above; that is,

$$ds^2 = d\rho^2 + (\sinh \rho)^2 (d\phi^2 + \sin^2 \phi d\theta^2). \qquad (9.3\text{-}4)$$

9.4 The Horosphere Again

From the line element derived in the last section, we now obtain a second proof that the geometry of a horosphere is Euclidean.

In Euclidean 3-space there is a surface (sphere) in which the angle sum of a triangle is greater than its Euclidean value π. That suggests the question, is there a surface in $\mathbb{H}^3$ in which the angle sum of a triangle is greater than its hyperbolic value (which is less than π) so as to be just equal to π? In other words, is there a surface whose geometry inherited from $\mathbb{H}^3$ is Euclidean? To say that the geometry of a surface is "inherited" from the 3-space in which it is embedded means that the line element ds of the surface is that obtained by simply restricting the line element ds of the 3-space to the surface.

The answer is that on an ordinary sphere in $\mathbb{H}^3$ the angle sum is greater than π, but it is equal to π on a horosphere (or "sphere of infinite radius").

Let ρ, ϕ, θ be spherical coordinates in $\mathbb{H}^3$, as before. In any plane $\theta =$ constant, ρ, ϕ are plane polar coordinates; hence, according to (6.1-2), the equation $\tanh(\rho/2) = \cos \phi$ $(0 < \phi \leq (\pi/2))$ gives half of a horocycle, as in Fig. 9.4. When that curve is rotated about the polar axis ℓ, the result is a horosphere. That is, the horosphere is the surface

$$\mathcal{S} : \tanh \frac{\rho}{2} = \cos \phi \qquad \left(0 < \phi \leq \frac{\pi}{2}, \text{ all } \theta\right). \qquad (9.4\text{-}1)$$

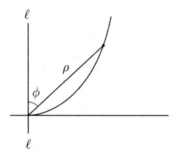

Fig. 9.4.

According to this equation, ρ varies from 0 to ∞ in $\mathcal{S}$ for each value of θ. It is clear that each horocycle in $\mathcal{S}$, obtained as the intersection of $\mathcal{S}$ with a plane $\theta = $ const, is a geodesic in $\mathcal{S}$. (See Exercise 3 below.) On that geodesic, we define a coordinate r as the distance along it from the origin. We wish to find r as a function of ρ. For $\theta = $ const $(d\theta = 0)$, we have

$$dr^2 = ds^2 = d\rho^2 + \sinh^2 \rho\, d\phi^2. \qquad (9.4\text{-}2)$$

We differentiate (9.4-1):

$$\frac{1}{2\cosh^2 \frac{\rho}{2}}\, d\rho = -\sin\phi\, d\phi.$$

In this equation we set $\sin\phi = \sqrt{1 - \cos^2\phi}$, where $\cos\phi$ is given by (9.4-1), so that $\sin\phi = (\cosh(\rho/2))^{-1}$; we solve the resulting equation for $d\phi$ and put the result into (9.4-2). The result is

$$dr^2 = \left(\cosh\frac{\rho}{2}\right)^2 d\rho^2.$$

[In (9.4-2) we have also used $\sinh\rho = 2\sinh(\rho/2)\cosh(\rho/2)$.] Hence, by integration,

$$r = \int \cosh\frac{\rho}{2}\, d\rho = 2\sinh\frac{\rho}{2}. \qquad (9.4\text{-}3)$$

From (9.3-4) in the preceding section, we then find for the line element in $\mathcal{S}$ in terms of the coordinate r, θ, after a bit of calculation,

$$ds^2 = dr^2 + r^2 d\theta^2. \qquad (9.4\text{-}4)$$

This is the line element of a Euclidean plane in polar coordinates.

Conclusion: The geometry that the horosphere inherits from $\mathbb{H}^3$ is Euclidean.

Exercises

1. Show that on an ordinary sphere in $\mathbb{H}^3$ the angle sum of a triangle is greater than π.

2. Define coordinates ξ, η, ζ in $\mathbb{H}^3$ in analogy with the coordinates ξ, η in $\mathbb{H}^2$ discussed in Section 3.9, in terms of an asymptotic bundle of lines and the family of horospheres orthogonal to them, so that holding η and ζ constant gives a line of the bundle, and keeping ξ constant gives a horosphere in which η and ζ are Cartesian coordinates (but with a length scale that depends on the value of ξ). What is the line element in those coordinates? What isometries are easily discussed in terms of those coordinates?

3. Show that a horocycle obtained by intersecting the horosphere $\mathcal{S}$ described above with a plane $\theta = $ const is a geodesic in $\mathcal{S}$, as claimed.

9.5 Models of $\mathbb{H}^3$

We state without proof that the system of axioms for $\mathbb{H}^3$ given in Chapter 4 is categorical; hence any model can be used to derive properties of hyperbolic 3-space. We describe two models. The first is the generalization of the Poincaré half-plane model. In it the parameters are x, y, z. The region Ω is the half-space

$$\Omega : z > 0, \quad \text{all } x, \quad \text{all } y.$$

The line element and the metric tensor are given by the generalizations of the formulas (7.3-2):

$$ds^2 = \frac{dx^2 + dy^2 + dz^2}{z^2} \qquad G = \left(\frac{1}{z}\right)^2 \begin{pmatrix} 1 & 0 & 0 \\ 0 & 1 & 0 \\ 0 & 0 & 1 \end{pmatrix}. \qquad (9.5\text{-}1)$$

The second model is the generalization of the Beltrami–Klein disk model. In it, the parameters X, Y, Z are the generalizations of the parameters X, Y of the disk model. They are given in terms of the spherical coordinates ρ, ϕ, θ by the equations that generalize (7.8-7):

$$\begin{aligned} X &= \tanh \rho \sin \phi \cos \theta, \\ Y &= \tanh \rho \sin \phi \sin \theta, \\ Z &= \tanh \rho \cos \phi. \end{aligned} \qquad (9.5\text{-}2)$$

The region Ω is the interior of the unit sphere

$$\Omega : X^2 + Y^2 + Z^2 < 1. \qquad (9.5\text{-}3)$$

In this model, the planes of $\mathbb{H}^3$ are represented by the disks obtained by intersecting Ω with planes $AX + BY + CZ = $ const, and the lines of $\mathbb{H}^3$ are represented by the chords of those disks.

Exercise 8 below deals with a model of n-dimensional hyperbolic space, namely, the analogue of the Poincaré half-plane model. The axioms for $\mathbb{H}^n$ are the obvious generalization of the axioms for $\mathbb{H}^3$ given in Chapter 4, and we state without proof that those axioms are also categorical.

Exercises

1. How are the lines and planes of $\mathbb{H}^3$ represented in the half-space model?
2. Verify the line element (9.5-1), for example, by showing that it gives the correct geometry in every plane that contains the Z-axis.
3. Describe the surfaces in $\mathbb{H}^3$ represented by planes $Z = $ const in the half-space model.
4. Show that each point of the plane $z = 0$ in the half-space model represents an ideal point at infinity of $\mathbb{H}^3$ and find the relation between the values of x, y in that plane and the values ϕ, θ of the spherical coordinates of that ideal point.
5. Show that the intersection of the Beltrami–Klein model with a plane that passes through the center of the sphere is a copy of the Beltrami–Klein disk model of $\mathbb{H}^2$ and from that that the lines of $\mathbb{H}^3$ are represented by chords of the sphere Ω. From that, conclude that the planes of $\mathbb{H}^3$ are represented by plane disks, as stated in the text.
6. From the Beltrami–Klein model, give a second proof of the fact stated in Chapter 4 that two nonintersecting planes can be asymptotic in at most one direction.
7. Find the formula for ds^2 in the Beltrami–Klein coordinates X, Y, Z. *Hint*: Rewrite the bracketed expression in (7.8-8) as $[dX^2 + dY^2 - (X dY - Y dX)^2]$ and generalize.
8. Define the hyperbolic space $\mathbb{H}^n$ as the space determined by the upper half-space of $\mathbb{R}^n$ given by coordinates $x_1, \ldots, x_n$, with x_n positive and the geometry determined by the Riemann metric given by

$$ ds^2 = \frac{dx_1{}^2 + \cdots + dx_n{}^2}{x_n^2}. $$

Describe the isometry group of this space in terms of a set of rotations and translations, following the method of Section 3.6. How many parameters does the isometry group have? Is the group simple? Reconcile the result with the earlier statement that the isometry group of $\mathbb{H}^3$ is isomorphic with $SL(2, \mathbb{C})$.
9. What is the isometry group of n-dimensional spherical space? Is it simple? Can it be expressed as a semi-direct product of simple groups? Do the same for $\mathbb{E}^n$.

9.6 Indefinite Metrics; Minkowski Geometry

Until now, we have assumed that the metric tensor G is a positive-definite matrix at each point of space (G is a function of the coordinates $u^1, \ldots, u^n$). That can be stated in either of two ways: first, an expression $v^T G v$ is positive for all nonzero choices of the vector v. Second, all the eigenvalues of G are positive. (Recall that G is symmetric, so that its eigenvalues are in any case real.) In the special and general theories of relativity one encounters geometries such that G is no longer positive-definite; the number of dimensions n is equal to 4 and there are three positive eigenvalues and one negative one (or, in some physics books one positive eigenvalue and three negative ones; the same theory can be formulated in either way). One says that then one is dealing with an *indefinite metric* and a *pseudo-Riemannian* manifold. The number of positive eigenvalues is necessarily the same throughout space, because G is required to be continuous and nonsingular, so that no eigenvalue can change sign from one point to another. (A nonsingular matrix is one that has an inverse; all its eigenvalues are $\neq 0$.) Furthermore, according to a theorem of linear algebra the number of positive eigenvalues is unchanged by a transformation of the form (5.2-10) from one coordinate system to another. (That theorem is known under the arcane name of Sylvester's law of inertia.)

The quadratic differential form

$$\sum_{j,k=1}^{n} g_{jk} du^j du^k$$

is now denoted by $\mathcal{F}(du^1, \ldots, du^n)$ or simply by $\mathcal{F}$ (it depends in general also on the coordinates $u^1, \ldots, u^n$ themselves). It was formerly denoted by ds^2, but, since it can be negative, that would suggest imaginary values of "ds."

The equation of a geodesic in the form (5.6-6), with the Christoffel symbols given by (5.6-4, 5) is still valid, but the interpretation is slightly different. The parameter on the geodesic, there denoted by s, cannot in general be interpreted as describing arclength. In fact, if u^4 is interpreted as time, and if the geodesic is timelike (see below), it is then interpreted as the motion of an object, and the parameter is then "proper time" — time as would be kept by a clock moving with the object. The quantity Φ given by (5.6-1) is still constant along a geodesic, but is not necessarily positive. For $\Phi > 0$, the geodesic is called *spacelike*; for $\Phi < 0$, it is *timelike*, and for $\Phi = 0$ it is a *null geodesic*. A spacelike geodesic may be thought of as a straight line present along its entire length at one instant of time; a timelike geodesic may be thought of as representing an object in uniform motion (or even stationary). A null geodesic may be thought of as representing a pulse of light (or a photon); along its path the proper time is "standing still." The concept of distance between points P and Q is lost; hence the triangle inequality is lost. The earlier notion of distance between P and Q along a path would be $\int |\Phi|^{1/2}ds$, but if the path consists of portions of null geodesics (which is always possible), the distance would be zero.

The case of indefinite metric to appear in the next chapter is that of *Minkowski geometry*, in which the metric tensor is

$$G = \begin{pmatrix} 1 & 0 & 0 & 0 \\ 0 & 1 & 0 & 0 \\ 0 & 0 & 1 & 0 \\ 0 & 0 & 0 & -1 \end{pmatrix}; \tag{9.6-1}$$

it is the geometry appropriate for special relativity.

Chapter 10

Connections with the Lorentz Group of Special Relativity

The Lorentz transformations and their origin in physics at the beginning of this century are described. It is shown that there is an isomorphism between the direct isometry group of the hyperbolic plane and the restricted Lorentz group in two space variables and time. A peculiar feature of the Lorentz group is shown to be connected with a property of the hyperbolic isometry group discussed in Section 8.3. We then generalize to one more dimension and show an isomorphism between the direct isometry group of three-dimensional hyperbolic space and the Lorentz group in three space variables and time. We then describe the so-called relativistic velocity space and show that it has the same geometry as the three-dimensional hyperbolic space.

10.1 Origin of Special Relativity

Maxwell's equations (1873) of the electromagnetic field in empty space are the following: $\mathbf{E}$ and $\mathbf{H}$ are 3-component vector fields, the *electric field* and the *magnetic field*, respectively, in certain standard units. They depend on x, y, z, t, and they satisfy the partial differential equations

$$\nabla \times \mathbf{E} + \frac{1}{c}\frac{\partial \mathbf{H}}{\partial t} = 0, \qquad \nabla \cdot \mathbf{H} = 0, \tag{10.1-1}$$

$$\nabla \times \mathbf{H} - \frac{1}{c}\frac{\partial \mathbf{E}}{\partial t} = 0, \qquad \nabla \cdot \mathbf{E} = 0, \tag{10.1-2}$$

where $\nabla \times$ denotes the curl of a vector field, and $\nabla \cdot$ denotes the divergence. (In a material medium, the two lower equations are modified by the presence, on the right side, of terms representing the density of electric current and the density of electric charge.) The equations were based on known laws of electrostatics, magnetostatics, and the law of induction, which says that a changing magnetic field can produce an electric field, together with Maxwell's hypothesis that a changing electric field can produce similarly a magnetic field. The constant c arose as the ratio of the units of charge in two systems of units; it is an absolute constant and has the dimensions of a speed (distance/time). Observation had shown that the speed of light is equal to c; hence, Maxwell's equations suggested that light consists of electromagnetic waves.

Prior to Maxwell's time all physics had been invariant under "Galilean" transformations from one frame of reference to a second frame moving at a constant velocity relative to the first, for example, under the transformation

$$x' = x - vt, \quad y' = y, \quad z' = z, \quad t' = t, \qquad (10.1\text{-}3)$$

where the constant v is the speed of the second frame with respect to the first. It came as quite a shock to discover that Maxwell's equations were not invariant under those transformations. Solutions of (10.1-1) and (10.1-2) can represent moving plane waves, moving at the speed c for any direction of propagation, but after the equations are transformed by (10.1-3), new terms appear in the equations, which have the effect that waves travel at different speeds in different directions. That seemed to indicate that there must be an absolute inertial frame of reference in the universe, namely, the unique one in which Maxwell's equations hold.

However, experiments like the Michelson–Morley experiment failed to detect any motion of a terrestrial laboratory through that special frame, for example, during the Earth's motion about the sun. It appeared that the speed of light is the same in all frames of reference; hence the relation between one frame and another could not be given by (10.1-3). That posed the problem of replacing (10.1-3) by a transformation law such that the speed of propagation of a light signal would be the same in both frames of reference, so that, for example, if x and t are related by $x - ct = \text{const}$, then, after the transformation, x' and t' would be related by $x' - ct' = \text{const}$. That led to what are now called *Lorentz transformations*, for example,

$$x' = \frac{x - vt}{\sqrt{1 - v^2/c^2}},$$
$$t' = \frac{t - vx/c^2}{\sqrt{1 - v^2/c^2}}, \qquad (10.1\text{-}4)$$
$$y' = y, \quad z' = z.$$

Lorentz transformations are discussed in the next section; they are the transformations under which the metric of Minkowski geometry, described in Section 9.6, is invariant. (From the above equations it is seen that if $x - ct$ is a constant, then $x' - ct'$ is a constant.)

The next problem was to find out how to transform Maxwell's equations so that electromagnetic theory would be fully invariant under Lorentz transformations. We discuss first the empty-space equations (10.1-1) and (10.1-2). Under rotations of the axes, without relative motion, **E** and **H** transform as vectors, and the question arises as to how they transform under a Lorentz transformation of the kind in (10.1-4). That can be done either by combining **E** and **H** in a skew-symmetric second-rank tensor or by expressing them in terms of a four-vector potential. We discuss the latter: it follows from Equations (10.1-1) that **E** and **H** can be expressed in terms of a vector potential **A** and a scalar potential ϕ by the equations

$$\mathbf{H} = \nabla \times \mathbf{A}, \qquad \mathbf{E} = -\nabla\phi - \frac{1}{c}\frac{\partial \mathbf{A}}{\partial t}, \qquad (10.1\text{-}5)$$

(Exercise 1 below), where the potentials **A** and ϕ satisfy the condition

$$\nabla \cdot \mathbf{A} + \frac{1}{c}\frac{\partial \phi}{\partial t} = 0 \qquad (10.1\text{-}6)$$

(Exercise 2 below), and it then follows from the second set of Maxwell's equations (10.1-2) that **A** and ϕ both satisfy the wave equation

$$\left(\nabla^2 - \frac{1}{c^2}\frac{\partial^2}{\partial t^2}\right)[\mathbf{A} \text{ or } \phi] = 0. \qquad (10.1\text{-}7)$$

Conversely, if **A** and ϕ satisfy (10.1-6) and (10.1-7), and if **E** and **H** are obtained from them by (10.1-5), then Maxwell's equations (10.1-1) and (10.1-2) hold (Exercise 3). The problem then arises, when the coordinates x, y, z, t are transformed by a Lorentz transformation (10.1-4), how must **A** and ϕ be transformed so that (10.1-6) and (10.1-7) continue to hold?

The three components of **A**, together with ϕ, are regarded as the components of a four-vector potential: $\Phi^j = A^j$ for $j = 1, 2, 3$, and $\Phi^4 = \phi$.

Then, if we introduce coordinates x^j by the equations

$$x^1 = x, \quad x^2 = y, \quad x^3 = z, \quad x^4 = ct, \qquad (10.1\text{-}8)$$

Equation (10.1-6) take the form

$$\sum_{j=1}^{4} \frac{\partial \Phi^j}{\partial x^j} = 0. \qquad (10.1\text{-}9)$$

According to the ideas of Section 5.10, this quantity is invariant (a scalar) if the Φ^j transform as the components of a contravariant vector, that is, according to (5.10-2). In terms of the coordinates (10.1-8), the Lorentz transformation takes the form

$$\begin{bmatrix} x^{1\prime} \\ x^{2\prime} \\ x^{3\prime} \\ x^{4\prime} \end{bmatrix} = \begin{bmatrix} b & 0 & 0 & a \\ 0 & 1 & 0 & 0 \\ 0 & 0 & 1 & 0 \\ a & 0 & 0 & b \end{bmatrix} \begin{bmatrix} x^1 \\ x^2 \\ x^3 \\ x^4 \end{bmatrix} \qquad (10.1\text{-}10)$$

where $b = \frac{1}{\sqrt{1-v^2/c^2}}$ and $a = bv/c$. In this case, the partial derivatives in (5.10-2) are all constant; hence the components of the four-vector potential Φ^j are transformed by the same linear equation, with the same matrix as above. Then, the quantity on the left of (10.1-9) is an invariant (a scalar), and the equation continues to hold in the primed variables (Exercise 4). The wave operator in (10.1-7) is also an invariant (Exercise 5). Hence, the fields $\mathbf{E}'$ and $\mathbf{H}'$ obtained from the $\Phi^{j'}$ satisfy Maxwell's equations in the primed variables. In the presence of matter, the current density and the charge density are combined into another four vector, which transforms in the same way as the four-vector potential. The entire electromagnetic theory is then invariant. (Perhaps the term "covariant" would be more appropriate, because vectors are involved.)

The difference between the fields $\mathbf{E}', \mathbf{H}'$ in the second frame of reference and those $\mathbf{E}, \mathbf{H}$ in the first obtained in this way is fully in agreement with observation. For example, in a frame moving through a frame containing a pure magnetic field, an electric field is observed (Exercise 6).

At the time of that work, very little was known about the structure of matter, but it was clear that electromagnetic phenomena must play a very dominant role; that led Einstein to conjecture, in 1905, that *all* physics is invariant under Lorentz transformations. That made it necessary to modify certain other parts of theoretical physics, for example, certain laws of mechanics, which Einstein then proceeded to do. That was the origin of special relativity. From that time on, there was no preferred frame of reference.

Exercises

1. Show that the fields $\mathbf{E}$ and $\mathbf{H}$ can be obtained from potentials as in (10.1-5). *Hint*: A divergence-free vector field can be expressed as the curl of another vector field, and a curl-free vector field can be expressed as the gradient of a scalar field.

2. Show that the potentials $\mathbf{A}$ and ϕ can be so chosen that Equation (10.1-6) holds. *Hint*: Show that if we add to $\mathbf{A}$ the gradient of an arbitrary scalar ψ and subtract from ϕ the time derivative of ψ divided by c, the fields $\mathbf{E}$ and $\mathbf{H}$ are unchanged, and ψ can be so chosen that Equation (10.1-6) holds. Show that it then follows from the second set of Maxwell's equations, (10.1-2), that $\mathbf{A}$ and ϕ both satisfy the wave equation (10.1-7).

3. Show that if the potentials satisfy (10.1-6) and (10.1-7) and the fields $\mathbf{E}$ and $\mathbf{H}$ are obtained from them by (10.1-5); then those fields satisfy Maxwell's equations.

4. Show that if the coordinates x^i and the four-vector potential Φ^i transform according to (10.1-10), then the quantity on the left of (10.1-9) is an invariant (scalar).

5. Show that under a Lorentz transformation (10.1-10) the wave operator that appears in (10.1-7) is invariant.

6. Find formulas for $\mathbf{E}', \mathbf{H}'$ in terms of $\mathbf{E}, \mathbf{H}$.

10.2 Lorentz and Poincaré Groups

A *Lorentz transformation* is a homogeneous linear transformation from variables x, y, z, t to variables x', y', z', t' such that the quadratic form Q defined by $Q(x, y, z, t) = t^2 - x^2 - y^2 - z^2$ is preserved,[3] that is, such that

$$t'^2 - x'^2 - y'^2 - z'^2 = t^2 - x^2 - y^2 - z^2 \qquad (10.2\text{-}1)$$

for all x, y, z, t. The generalization to any finite number of space variables and time is evident [and also, to any finite number of space variables and any finite number of time variables; in the next section, we shall be concerned with the case of two space variables and time (x, y, t)].

We use vector–matrix notation and write

$$\begin{pmatrix} x' \\ y^{\iota} \\ z' \\ t' \end{pmatrix} = L \begin{pmatrix} x \\ y \\ z \\ t \end{pmatrix}, \qquad (10.2\text{-}2)$$

where L is a 4×4 real matrix, called a *Lorentz matrix*. If we define

$$J = \begin{pmatrix} -1 & 0 & 0 & 0 \\ 0 & -1 & 0 & 0 \\ 0 & 0 & -1 & 0 \\ 0 & 0 & 0 & 1 \end{pmatrix}, \qquad (10.2\text{-}3)$$

then the quadratic form is

$$(x \ y \ z \ t) J \begin{pmatrix} x \\ y \\ z \\ t \end{pmatrix}, \qquad (10.2\text{-}4)$$

hence L is a Lorentz matrix if and only if

$$L^T J L = J. \qquad (10.2\text{-}5)$$

Clearly, the inverse of such a transformation and the composition of two such transformations also preserve the quadratic form Q; hence the set of all Lorentz transformations is a non-Abelian continuous group.

In 1905 it was known, through the work of Lorentz, Fitzgerald, Poincaré, and others, that Maxwell's electromagnetic theory was invariant under Lorentz transformations (not under the classical transformations of relative motion without change of the time variable). Since electromagnetism appeared to be a very substantial part of physics, Einstein conjectured in 1905 that all physics was also invariant under Lorentz transformations. (That made it necessary for Einstein to modify the classical mechanics

---

[3] We assume units such that the speed of light, c, is 1.

of the motion of particles and their collisions.) The validity of Einstein's conjecture was subsequently verified by experiments and observations in all branches of physics; thus was special relativity created.

Since the origin of the space variables and the origin of time are clearly irrelevant, it is clear that physical laws are also invariant under the non-homogeneous transformations

$$\begin{pmatrix} x' \\ y' \\ z' \\ t' \end{pmatrix} = L \begin{pmatrix} x \\ y \\ z \\ t \end{pmatrix} + \begin{pmatrix} a \\ b \\ c \\ d \end{pmatrix}, \qquad (10.2\text{-}6)$$

where a, b, c, d are real constants. These transformations are called Poincaré transformations, and the resulting group is called the Poincaré group. In the following, we shall be concerned only with the homogeneous case; hence with the Lorentz group.

By taking determinants in (10.2-5), it is seen that $\det L = \pm 1$. Furthermore, if we take $x = y = z = 0$ in (10.2-2), we have $t' = L_{44}t$; hence, from (10.2-1), we see that L_{44} must be either ≥ 1 or ≤ -1. The transformations with $\det L = 1$ and $L_{44} \geq +1$ constitute the *restricted* Lorentz group; the other three cases are transformations with spatial inversion or time reversal or both. We shall be concerned only with the restricted Lorentz group.

It is easily seen that if a and b are constants such that $a^2 - b^2 = 1$, and a is > 0, the matrix

$$\begin{pmatrix} a & 0 & 0 & b \\ 0 & 1 & 0 & 0 \\ 0 & 0 & 1 & 0 \\ b & 0 & 0 & a \end{pmatrix}$$

is a restricted Lorentz matrix. We set $X = \sinh^{-1} b$ (then, $a = \cosh X$) and we write

$$L_X = \begin{pmatrix} \cosh X & 0 & 0 & \sinh X \\ 0 & 1 & 0 & 0 \\ 0 & 0 & 1 & 0 \\ \sinh X & 0 & 0 & \cosh X \end{pmatrix}. \qquad (10.2\text{-}7)$$

Also, if M is any rotation matrix (a 3×3 real orthogonal matrix of determinant 1), then the matrix

$$R = \begin{pmatrix} & & & 0 \\ & M & & 0 \\ & & & 0 \\ 0 & 0 & 0 & 1 \end{pmatrix} \qquad (10.2\text{-}8)$$

is also a restricted Lorentz matrix. It can be proved that the most general restricted Lorentz matrix can be written as $R'L_X R$ for a suitable value of X and suitable rotations R and R'. (See the Exercise below.)

Exercise

1. Show that any restricted Lorentz matrix L can be written in the form $L = R'L_X R$. *Hint*: First determine X from L_{44}.

10.3 Isomorphism of the Restricted Lorentz Group in Two Space Variables and Time with the Direct Isometry Group of the Hyperbolic Plane

We recall that two groups are isomorphic if there is a one-to-one mapping from the elements of the first group to those of the second which preserves the group operations, that is, maps products to products and inverses to inverses. Then, the two groups, regarded abstractly, are identical; they have the same intrinsic structure.

We shall demonstrate the isomorphism in two ways:

(1) We shall exhibit explicitly an isomorphism between the group of Lorentz matrices and the group of fractional–linear transformations of the Poincaré half-plane model which represent the isometries of the hyperbolic plane.

(2) We shall describe a surface $\mathcal{S}$ in the three-dimensional space $\mathbb{R}^3$ of x, y, t which is a *model* of the hyperbolic plane. That is, we shall exhibit a one-to-one mapping from $\mathcal{S}$ to $\mathbb{H}^2$ in terms of coordinates ρ, θ in $\mathcal{S}$ which are set equal to polar coordinates ρ, θ in $\mathbb{H}^2$ (about some origin), and we shall show that, in these variables, the differential form $\mathcal{F} = dx^2 + dy^2 - dt^2$ that is invariant in $\mathcal{S}$ under the Lorentz transformations in Minkowski space is identical with the line elements in $\mathbb{H}^2$ that is invariant under the isometries of $\mathbb{H}^2$. Since all geometrical concepts (distances, lines or geodesics, angles, and so on) can be obtained from the line element, we see that $\mathcal{S}$ is a model of $\mathbb{H}^2$.

First, we recall that, according to Section 7.4, there is a one-to-one correspondence between the isometries of $\mathbb{H}^2$ and the fractional-linear transformations $z \to z' = (az + b)/(cz + d)$ $(a, b, c, d$ real) of the Poincaré half-plane model of $\mathbb{H}^2$. We represent each such transformation by the matrix $P = \begin{pmatrix} a & b \\ c & d \end{pmatrix}$, where we assume that $\det P = 1$, that is, $ad - bc = 1$; then there is a one-to-two correspondence between isometries and matrices, P and $-P$ giving the same fractional–linear transformation, hence the same isometry. Compositions of isometries are obtained by the products of the corresponding matrices.

We now let such a 2×2 matrix P be given, and we define a transformation from x, y, t to x', y', t' by the equation

$$\begin{pmatrix} t' - y' & x' \\ x' & t' + y' \end{pmatrix} = P \begin{pmatrix} t - y & x \\ x & t + y \end{pmatrix} P^T. \tag{10.3-1}$$

To see that that is always possible, we note that for any x, y, t, the matrix $\left(\begin{smallmatrix} t-y & x \\ x & t+y \end{smallmatrix} \right)$ is symmetric. Since the transpose of a product of matrices is also equal to the product of the transposes in the opposite order, we see that the product on the right of the above equation is a symmetric matrix; hence it can be written in the form on the left, for suitable values of x', y', t'.

For a fixed matrix P, the transformation from x, y, t to x', y', t' is linear and homogeneous, because if M_1 and M_2 are any matrices of the form $\left(\begin{smallmatrix} t-y & x \\ x & t+y \end{smallmatrix} \right)$, and M_1' and M_2' are their transforms according to (10.3-1) and if a and b are constants, then $aM_1' + bM_2' = P(aM_1 + bM_2)P^T$. Also, the zero matrix (where $x = y = t = 0$) is mapped onto the zero matrix. Furthermore, by taking determinants in (10.3-1) and recalling that $\det P = 1$, we see that

$$t'^2 - x'^2 - y'^2 = t^2 - x^2 - y^2. \qquad (10.3\text{-}2)$$

Therefore, (10.3-1) determines a Lorentz transformation hence a Lorentz matrix L, and we write $L = \sigma(P)$; the proof that it is a restricted Lorentz transformation is left to the reader.

We now study the mapping $\sigma : P \to L = \sigma(P)$. If we transform from x, y, t by P as in (10.3-1) and then transform from x', y', t' to x'', y'', t'' by Q in the same manner, the result is the same as transforming from x, y, t to x'', y'', t'' by the product QP; that is, $\sigma(QP) = \sigma(Q)\sigma(P)$, so σ maps products onto products. Since P and $-P$ give the same Lorentz transformation and also the same isometry of $\mathbb{H}^2$, to show the one-to-one relation between Lorentz transformations and the isometries, we must show that a given Lorentz transformation is given only by a single matrix P and its negative. First, if $L = \sigma(P)$ is the identity Lorentz transformation, we must have $A = PAP^T$ for every symmetric matrix A. By taking A first as $\left(\begin{smallmatrix} 1 & 0 \\ 0 & 0 \end{smallmatrix} \right)$, then as $\left(\begin{smallmatrix} 0 & 1 \\ 1 & 0 \end{smallmatrix} \right)$, and then as $\left(\begin{smallmatrix} 0 & 0 \\ 0 & 1 \end{smallmatrix} \right)$, it is easy to see that P has to be either $\pm I$, where I is the unit matrix $\left(\begin{smallmatrix} 1 & 0 \\ 0 & 1 \end{smallmatrix} \right)$. It then follows that a given Lorentz matrix L cannot be given by two different pairs $\pm P$ and $\pm Q$, for then the identity matrix LL^{-1} would be given by PQ^{-1}, which would not be of the required form $\pm I$.

We have found an isomorphism from the isometry group of $\mathbb{H}^2$ and *some* group of Lorentz transformations, and what remains is to show that *all* restricted Lorentz transformations are obtained in this way. We consider special cases. First, let P be the matrix

$$P = \begin{pmatrix} \cosh(X/2) & \sinh(X/2) \\ \sinh(X/2) & \cosh(X/2) \end{pmatrix}.$$

When the matrix product on the right of (10.3-1) is multiplied out, a little calculation shows that

$$x' = (\cosh X)x + (\sinh X)t$$
$$y' = y$$
$$t' = (\sinh X)x + (\cosh X)t.$$

In this case, the Lorentz matrix is the 3×3 version of (10.2-7). Second, we take

$$P = \begin{pmatrix} \cos(\alpha/2) & \sin(\alpha/2) \\ -\sin(\alpha/2) & \cos(\alpha/2) \end{pmatrix}.$$

In this case, the calculation gives

$$x' = (\cos\alpha)x - (\sin\alpha)y,$$
$$y' = (\sin\alpha)x + (\cos\alpha)y,$$
$$t' = t.$$

Hence, the Lorentz matrix is of the form (10.2-8), where M is now the 2×2 rotation matrix $\left(\begin{smallmatrix} \cos\alpha & -\sin\alpha \\ \sin\alpha & \cos\alpha \end{smallmatrix} \right)$. According to Exercise 1 in Section 10.2, the general restricted Lorentz transformation is a composition of transformations of these two types. This completes the proof of the isomorphism.

We now define a surface $\mathcal{S}$ (hyperboloid of revolution) in Minkowski space by the equations

$$\mathcal{S} : t^2 - x^2 - y^2 = 1 \qquad t \geq 1. \tag{10.3-3}$$

According to (10.3-2), $\mathcal{S}$ is invariant (mapped onto itself) under the Lorentz transformations. We define coordinates ρ, θ in $\mathcal{S}$ by the equations

$$x = \sinh\rho\cos\theta,$$
$$y = \sinh\rho\sin\theta, \qquad \rho \geq 0,\ 0 \leq \theta < 2\pi. \tag{10.3-4}$$
$$t = \cosh\rho.$$

The first two of these equations can be taken as determining ρ, θ, and then the third equation follows from (10.3-3).

The differential form $dx^2 + dy^2 - dt^2$ of the Minkowski space is invariant under all Lorentz transformations. A short calculation shows that, when this form is restricted to the surface $\mathcal{S}$, it takes the form

$$dx^2 + dy^2 - dt^2 = d\rho^2 + (\sinh\rho)^2 d\theta^2. \tag{10.3-5}$$

According to Chapter 7, this is precisely the invariant differential line element of the polar-coordinate model of $\mathbb{H}^2$. According to the comments in paragraph (2) at the beginning of this section, that proves that $\mathcal{S}$ is a model of $\mathbb{H}^2$. Each Lorentz transformation leads to an isometry, and conversely, each isometry leads, via (10.3-4), to a Lorentz transformation.

10.4 A Pseudoparadoxical Feature of the Lorentz Group

The effect of the Lorentz transformation that corresponds to the Lorentz matrix (10.2-7) is that the x', y', z' frame of reference is moving with a certain velocity in the x direction relative to the x, y, z frame. However, the new x-, y-, and z-axes are parallel to the old ones; there is only a relative velocity with no rotation. Such a Lorentz transformation is called a "boost" by the physicists. A boost with relative velocity in a general direction is given by a matrix of the form $RL_X R^T$, hence is always a symmetric matrix. However, the product of two symmetric matrices is in general not symmetric. Therefore, if P and Q are symmetric, hence represent boosts, their resultant QP is in general not a boost. In any case, however, QP can be written as $R'T_X R$, for a suitable value of X and suitable rotations R' and R, as stated at the end of Section 10.2. If we write the product QP as $(R'R)(R^T T_X R)$ (note that RR^T is the unit matrix, because R is orthogonal), then $R^T T_X R$ is another boost, and $(R'R)$ is a rotation. We conclude: *the resultant of three pure boosts can be a pure rotation.* In the early days of quantum mechanics, that effect was shown by L. H. Thomas to produce a relativistic correction to certain atomic energy levels, because an electron orbiting a nucleus undergoes in effect a succession of small boosts, and their resultant has an effect similar to electron spin. This feature of the Lorentz group is related to the feature of the hyperbolic isometry group mentioned in Chapter 3, namely, that the resultant of three pure translations can be a rotation in the plane.

10.5 Generalization to Three Space Variables and Time

We now consider Lorentz transformations in the Minkowski space from x, y, z, t to x', y', z', t' as in Section 10.1. In $\mathbb{R}^4$ we define a hypersurface $\mathcal{S}$ as

$$\mathcal{S} : t^2 - x^2 - y^2 - x^2 = 1 \qquad (t \geq 1). \qquad (10.5\text{-}1)$$

It is invariant (transformed into itself) under all restricted Lorentz transformations. We define coordinates ρ, ϕ, θ in $\mathcal{S}$ by the equations

$$
\begin{aligned}
x &= \sinh \rho \sin \phi \cos \theta, \\
y &= \sinh \rho \sin \phi \sin \theta, \\
z &= \sinh \rho \cos \phi, \\
t &= \cosh \rho.
\end{aligned}
\qquad (10.5\text{-}2)
$$

In analogy with the previous case, the first three of these equations can serve to define the coordinates ρ, ϕ, θ, and then the fourth equation is satisfied automatically because of (10.5-1).

The differential form $dx^2 + dy^2 + dz^2 - dt^2$ is invariant under all Lorentz transformations, and, when restricted to the hypersurface, is given by

$$dx^2 + dy^2 + dz^2 - dt^2 = d\rho^2 + \sinh^2 \rho (d\phi^2 + \sin^2 \phi d\theta^2). \qquad (10.5\text{-}3)$$

According to Section 9.5, the expression on the right is the line element of $\mathbb{H}^3$ in spherical coordinates. Therefore, the inherited geometry of the hypersurface $\mathcal{S}$ is three-dimensional hyperbolic geometry.

10.6 Relativistic Velocity Space

We shall describe just enough of the physics of special relativity to introduce the *velocity space* for moving particles and show that it is isomorphic to the hyperbolic space $\mathbb{H}^3$.

By a *frame of reference* we mean a "laboratory" in which Cartesian coordinates x, y, z are measured in the usual way by rulers and the time t by a clock. The idea of special relativity is that if a second frame of reference, with coordinates x', y', z' and time t', is in uniform motion with respect to the first, then x', y', z', t' are related to x, y, z, t by a Poincaré transformation.

If a particle is in motion through the x, y, z, t frame of reference, it traces out a curve or path in that four-dimensional space, and we describe that path by writing

$$x = x(\tau), \quad y = y(\tau), \quad z = z(\tau), \quad t = t(\tau),$$

where τ is a parameter. It turns out to be convenient to choose τ as determined along the path by the equation $d\tau^2 = dt^2 - dx^2 - dy^2 - dz^2$, that is, after dividing by $d\tau^2$, by the equation

$$1 = \left(\frac{dt}{d\tau}\right)^2 - \left(\frac{dx}{d\tau}\right)^2 - \left(\frac{dy}{d\tau}\right)^2 - \left(\frac{dz}{d\tau}\right)^2.$$

[That that is always possible results from a physical principle, according to which the magnitude of a particle's velocity (speed) is always less than the speed of light, which we have taken to be 1 by suitable choice of units. The velocity components, in the usual sense, are

$$\frac{dx}{d\tau}\bigg/\frac{dt}{d\tau}, \quad \frac{dy}{d\tau}\bigg/\frac{dt}{d\tau}, \quad \frac{dz}{d\tau}\bigg/\frac{dt}{d\tau},$$

hence, along the path, $dt^2 - dx^2 - dy^2 - dz^2$ is always positive.]

The 4-vector whose components are $u^j(\tau)$ $(j = 0, 1, 2, 3)$, where

$$u^0 = \frac{dt}{d\tau}, \quad u^1 = \frac{dx}{d\tau}, \quad u^2 = \frac{dy}{d\tau}, \quad u^3 = \frac{dz}{d\tau},$$

is called the *velocity 4-vector*; it satisfies the identity

$$1 = {u^0}^2 - {u^1}^2 - {u^2}^2 - {u^3}^2 \tag{10.6-1}$$

at all points of the path.

If we view the same motion of that particle in a different frame of reference, with coordinates and time x', y', z', t' obtained from x, y, z, t by a Lorentz transformation (10.2-2), with the same parameter τ along the path, then,

$$
\begin{pmatrix} \dfrac{dx'}{d\tau} \\ \dfrac{dy'}{d\tau} \\ \dfrac{dz'}{d\tau} \\ \dfrac{dt'}{d\tau} \end{pmatrix} = L \begin{pmatrix} \dfrac{dx}{d\tau} \\ \dfrac{dy}{d\tau} \\ \dfrac{dz}{d\tau} \\ \dfrac{dt}{d\tau} \end{pmatrix} ;
$$

hence the new 4-vector velocity components $u^{1'}, u^{2'}, u^{3'}, u^{0'}$ are obtained from the old ones by the same Lorentz transformation as x, y, z, t, namely,

$$
\begin{pmatrix} u^{1'} \\ u^{2'} \\ u^{3'} \\ u^{0'} \end{pmatrix} = L \begin{pmatrix} u^{1} \\ u^{2} \\ u^{3} \\ u^{0} \end{pmatrix} . \tag{10.6-2}
$$

Equation (10.6-1) is invariant under that transformation.

We now define a geometry in the space $\{u^1, u^2, u^3, u^0\}$ by defining a differential form $\mathcal{F}'$ in analogy with the form $\mathcal{F}$ in the space $\{x, y, z, t\}$; that is,

$$
\mathcal{F}' = (du^1)^2 + (du^2)^2 + (du^3)^2 - (du^0)^2.
$$

Then, according to the arguments given in the preceding Sections 10.3–10.5, the hypersurface $\mathcal{S}$ determined by the restriction (10.6-1), which is called the *relativity velocity space*, is isomorphic with the hyperbolic space $\mathbb{H}^3$. It is transformed into itself by (10.6-2), and its isometry group is isomorphic with the Lorentz group in three space variables and time.

Chapter 11

Constructions by Straightedge and Compass in the Hyperbolic Plane

As in the Euclidean case, questions of constructibility by straightedge and compass involve consideration of certain algebraic number fields called quadratic-surd fields. They are obtained by starting with the rational field and adjoining square roots in succession. Roughly speaking, a point with polar coordinates r, θ can be constructed in the Euclidean case if r and $\sin \theta$ are in such a field (in which case $\cos \theta$ and $\tan \theta$ are also in such a field), while in the hyperbolic case, the point can be constructed if $\tanh r$ and $\sin \theta$ are in such a field. The first two sections give the proof (which is not short) that if the point can be constructed, then $\tanh r$ and $\sin \theta$ are in a quadratic-surd field. Section 11.6 contains the proof (also not short) that every such point can be constructed. As a result, there are certain differences between Euclidean and the hyperbolic cases. In the hyperbolic case, a segment cannot in general be trisected, while some circles can be squared (that depends on the radius). In the general theory, the constructions start with a single given line and a single given point on that line, but no segment of unit length, because a unit of length is implied by the hyperbolic axioms. Certain construction problems are considered in which other things are given in advance, for example constructing the angle of parallelism $\Pi(y)$, when a segment of length y is given. Since any two ideal points in the hyperbolic plane determine a unique line, it is assumed that the straightedge can draw the line between any two ideal points, but it is shown in Section 11.9 that that is not really necessary. All constructions can be made (with more trouble) if the straightedge is only assumed to be able to draw a line between any two finite points. The purpose of the last section is to show that constructibility is not merely a matter of interest for engineering or draftsmanship, but also throws light on the geometry itself. The set of constructible points (it is a countable set!) constitutes a geometry that satisfies all the axioms of the hyperbolic plane, except that there is no axiom of completeness, either for the real number system $\mathbb{R}$ or for the points on a line. (Those axioms are of course not categorical, since, in particular, the ordinary hyperbolic plane satisfies them.)

11.1 Definitions and Examples; Quadratic-Surd Fields

Operations by straightedge and compass are understood to be the same as in the Euclidean case, except for the following two differences:

(a) In the hyperbolic plane, any two points determine a unique line, even if one or both of the points is an ideal point at infinity. Therefore, a "straightedge," it will be assumed, can draw a line between *any* two points, finite or ideal. (This is mathematics, not physics; we are dealing with abstract concepts, not finite solid objects.) It will be shown in Section 11.9 that this assumption is not really necessary, but it simplifies the discussion.

(b) In the hyperbolic plane, it is not necessary for a segment of unit length (or of any known length) to be given in advance, because a length scale is inherent in the geometry. It is nevertheless necessary to *define* a unit of length (see Section 1.2). That is done by assigning the value 1 to the constant K that appeared in Chapter 6 (and already in Section 3.6). Then, the lengths of certain constructible segments can be determined by the formulas of Chapter 6 in terms of the unit so defined. (Curiously enough, it is not possible to construct a segment of unit length.)

We assume a knowledge of the following constructions, which are the same as in the Euclidean case: (1) erecting a perpendicular to a line at a point of that line, (1) dropping a perpendicular from a point to a line, (3) bisecting a segment, (4) adding or subtracting segments, (5) bisecting an angle, and (5) adding or subtracting angles. In each of these, certain points and lines are given; in some of them, an additional segment has to be chosen arbitrarily and used in the construction but does not appear in the final result, which is independent of that choice. See exercises below. (*Note*: We obviously cannot include constructions that depend on drawing parallels.)

In most of this chapter, the initial configuration for a construction will consist of a single point Z and a single line ℓ passing through Z. (In particular, we are not given a unit segment or a segment of known length. In Sections 11.4, 11.5, and 11.7, a different kind of construction problem is considered, in which further things are given in advance.)

The following construction will give us a first segment of known length. We construct a line m perpendicular to ℓ at Z; let L be the ideal point at one end of ℓ and M the ideal point at one end of m. With the straightedge we draw the line n joining the ideal points L and M, and then we drop a perpendicular from Z to the line n, as indicated schematically in Fig. 11.1a, and we let P be the foot of the perpendicular. From the isometries of the plane, it is seen that if the procedure is repeated starting with any other line ℓ' and point Z' on it, then the resulting segment $Z'P'$ is congruent to ZP, hence the length $|ZP|$ has been determined in a way that does not depend on any arbitrary choice. See Section 11.9 for a method of performing this construction with a "finite" straightedge.

The length $|ZP|$ constructed in this way might be taken as the unit of length, but the following procedure is slightly more convenient. The

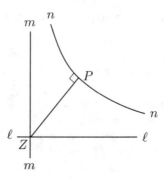

Fig. 11.1a.

angle of parallelism $\Pi(y)$ for distance y is given by formula (6.1-1), namely, $e^{-Ky} = \tan(\Pi(y)/2)$, where K is a constant. We define the unit of length to be such that $K = 1$. In terms of that unit, the length $y = |ZP|$ is given by the equation $e^{-y} = \tan(\pi/8) = \sqrt{2} - 1$, hence by $|ZP| = \ln(\sqrt{2} + 1)$, so that $\sinh |ZP| = 1$. We shall generalize this construction presently.

To simplify the discussion in the next section, we adopt the convention that, after the above preliminary construction, whenever an arbitrary length is required, it is taken from the figure that then exists. That is always possible. For erecting a perpendicular to a line at a point of the line, an arbitrary radius is needed, but it can be *any* radius, hence might as well be taken from a segment in the figure. The same applies to bisecting an angle (if it is done correctly). Bisecting a segment requires no other segment. Dropping a perpendicular from a point to a line requires a radius at least as large as the distance from the point to the line, which is not available until the construction is complete. In this case, we take a length d, construct multiples $2d, 3d$, etc., by additions of segments on a line. In a finite number of trials (none of which requires an "arbitrary" segment) we obtain a circle about the given point that intersects the given line, and so on.

By successive angle bisections, starting with a right angle, and then by angle additions, one can construct an angle of the form

$$\alpha = \pi \frac{\text{integer}}{\text{power of 2}}. \tag{11.1-1}$$

We consider first the case in which the integer is 1 and we call

$$s_n = \cot 2^{-n} \pi.$$

From the law for angle doubling, we can find a relation between s_{n-1} and s_n, and that gives the quadratic equation

$$(s_n^2 - 1) - 2s_{n-1}s_n = 0$$

for s_n, from which we can determine the s_n's inductively by

$$s_2 = \cot \frac{\pi}{4} = 1$$
$$s_n = s_{n-1} + \sqrt{s_{n-1}^2 + 1} \qquad\qquad (11.1\text{-}2)$$

The quantities s_n are members of some so-called quadratic-surd number field. Such fields will now be defined. We recall from the Appendix to Chapter 1 that a *field* $\mathbb{F}$ is a set of quantities that is closed under addition, subtraction, multiplication, and division by nonzero elements. The field of rational numbers is denoted by $\mathbb{Q}$. If α is a quantity not in a field $\mathbb{F}$, we denote by $\mathbb{F}(\alpha)$ the smallest field that contains all elements of $\mathbb{F}$ and α. It contains all rational functions of α with coefficients in $\mathbb{F}$. We say that α has been *adjoined* to the field $\mathbb{F}$. In case α satisfies an algebraic equation

$$a_n \alpha^n + a_{n-1}\alpha^{n-1} + \cdots + a_1\alpha + a_0 = 0, \qquad (11.1\text{-}3)$$

where the coefficients a_i are in $\mathbb{F}$, then $\mathbb{F}(\alpha)$ is an *algebraic extension* of $\mathbb{F}$, and if the above equation is irreducible (i.e., the polynomial on the left cannot be written as the product of two or more nonconstant polynomials with coefficients in $\mathbb{F}$), then n is the *degree* of the extension or the *degree* of $\mathbb{F}(\alpha)$ *over* $\mathbb{F}$.

We now state without proof a few facts about field extensions. For more detail, see *Modern Algebra* by B. L. van der Waerden, Frederick Ungar, 1949, or *A Survey of Modern Algebra* by G. Birkhoff and S. MacLane, MacMillan, 1953, or do the exercises at the end of this section.

(1) If n is the degree of $\mathbb{F}(\alpha)$ over $\mathbb{F}$, then any element x of $\mathbb{F}(\alpha)$ can be written uniquely in the form

$$x = b_0 + b_1\alpha + \cdots + b_{n-1}\alpha^{n-1}, \qquad (11.1\text{-}4)$$

where the coefficients b_i are in $\mathbb{F}$. See Exercises 1, 2, and 3 below.

(2) If β is not in $\mathbb{F}(\alpha)$, then a further extension, denoted either by $\mathbb{F}(\alpha)(\beta)$ or by $\mathbb{F}(\alpha, \beta)$, can be obtained by adjoining β to $\mathbb{F}(\alpha)$. The result of any finite number of extensions $\mathbb{F}(\alpha)(\beta)(\gamma)\cdots$, where each of $\alpha, \beta, \gamma \cdots$ is algebraic over the field to which it is adjoined, is a *finite algebraic extension* of $\mathbb{F}$.

(3) (Theorem of the primitive element) If $\mathbb{K}$ is any finite algebraic extension of $\mathbb{F}$, then there exists an element θ of $\mathbb{K}$ such that $\mathbb{K} = \mathbb{F}(\theta)$; the degree of the element θ is the *degree* of the extension. See example in Exercise 4.

(4) If n is the degree of $\mathbb{F}(\alpha)$ over $\mathbb{F}$ and m is the degree of $\mathbb{F}(\alpha)(\beta)$ over $\mathbb{F}(\alpha)$, then the degree of $\mathbb{F}(\alpha)(\beta)$ over $\mathbb{F}$ is the product nm. See Exercise 6.

Definition 11.1: A *quadratic-surd* field is a real number field $\mathbb{F}_k$ that can be defined by writing

$$\mathbb{F}_0 = \mathbb{Q} \quad \text{(the field of real rational numbers)},$$
$$\mathbb{F}_i = \mathbb{F}_{i-1}(\sqrt{w_i}) \quad (i = 1, 2, \dots, k), \tag{11.1-5}$$

where, for each i, the quantity $\sqrt{w_i}$ adjoined is the square root of a positive quantity w_i in $\mathbb{F}_{i-1}$, and $\sqrt{w_i}$ is not itself in $\mathbb{F}_{i-1}$.

By repeated use of item (4) above, we see that the degree of $\mathbb{F}_k$ over $\mathbb{Q}$ is 2^k. As a further application of item (4), we see that if x is the root of an irreducible cubic equation with rational coefficients, then x is not in a quadratic-surd field, because 3 does not divide any power of 2.

We see that each of the quantities s_n defined by (11.1-2) is in a quadratic-surd field, hence so is the tangent of $2^{-n}\pi$, which is $1/s_n$. More generally, call

$$t_{n,p} = \tan(2^{-n}p\pi),$$

where p is a positive integer less than 2^n. From the formula for the tangent of the sum of two angles, we find that

$$t_{n,p+1} = \frac{t_{n,p} + t_{n,1}}{1 - t_{n,p}t_{n,1}};$$

hence all the $t_{n,p}$ are in a quadratic-surd field.

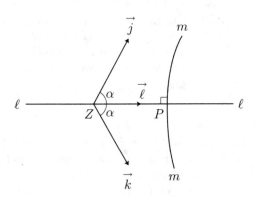

Fig. 11.1b.

Now let α be an angle of the form (11.1-1) between 0 and $\pi/2$, let $\vec{j}$ and $\vec{k}$ be rays from Z, one in each half-plane bounded by ℓ, each at an angle α with respect to a ray $\vec{\ell}$ along ℓ from Z, as in Fig. 11.1b. Let J be the ideal point at the end of $\vec{j}$ and K that at the end of $\vec{k}$. By the straightedge, we construct the line m from J to K. Then, it is easily seen, m is perpendicular to ℓ, and, according to (6.1-1), the distance $y = |ZP|$

from Z to m satisfies the equation $e^{-y} = \tan \dfrac{\alpha}{2}$. For a length y that can be constructed in this way, e^{-y} is in a quadratic-surd field.

Exercises

1. Show that if $f(x)$ is any rational function with rational coefficients, that is, $f(x) = (p(x))/(q(x))$, where $p(x)$ and $q(x)$ are polynomials with rational coefficients, then $f(\sqrt{2})$ can always be written as $a + b\sqrt{2}$, where a and b are rational. This is a special case of (11.1-4) with $n = 2$ and $\mathbb{F} = \mathbb{Q}$.

2. The first step in deriving (11.1-4) more generally is to show that if α satisfies equation (11.1-3), where the a_i are in $\mathbb{F}$, then any polynomial in α with coefficients in $\mathbb{F}$ can be written as (i.e., is equal to) a polynomial in α of degree $\leq n - 1$, with coefficients in $\mathbb{F}$. *Hint:* The higher powers can be expressed in terms of lower powers by use of (11.1-3).

3. From the preceding exercise, we see that any rational function $f(\alpha)$ can be written as $(p(\alpha))/(q(\alpha))$, where $p(x)$ and $q(x)$ are of degree $\leq n - 1$. Now we have to "get rid" of the denominator, that is, replace it by an element of $\mathbb{F}$. Show by use of the Euclidean algorithm for polynomials that if $r_0(x)$ and $r_1(x)$ are any polynomials with coefficients in $\mathbb{F}$, then their greatest common divisor (GCD) can be expressed as $d(x) = r_0(x)R_0(x) + r_1(x)R_1(x)$, where $R_0(x)$ and $R_1(x)$ are further polynomials with coefficients in $\mathbb{F}$. Show that as follows. Assume that the degree of r_0 is $\geq$ the degree of r_1 (otherwise, interchange r_0 and r_1). Divide r_0 by r_1 to give

$$r_0(x) = q_1(x)r_1(x) + r_2(x),$$

where $q_1(x)$ is the quotient and $r_2(x)$ is the remainder. The degree of r_2 is less than that of r_1. Now divide r_1 by r_2 to give

$$r_1(x) = q_2(x)r_2(x) + r_3(x),$$

where the degree of r_3 is less than that of r_2. Continue this procedure until a division comes out even, so that $r_{k+1}(x)$ is identically zero, for some k. Show from these equations that $r_k(x)$ divides each of the preceding remainders, so that, in particular, it is a common divisor of $r_0(x)$ and $r_1(x)$. Then, to show that it is the *greatest* common divisor, proceed as follows. The next-to-last equation expresses r_k as a linear function of r_{k-1} and r_{k-2} with coefficients that are polynomials with coefficients in $\mathbb{F}$. Then, by use of the equation before that, one can express r_k as a similar linear function of r_{k-2} and r_{k-3}. If we continue this process, we find eventually r_k expressed as a linear function of r_0 and r_1, say $r_0(x)R_0(x) + r_1(x)R_1(x)$, and it follows that any polynomial that divides both r_0 and r_1 divides r_k, so that $r_k(x)$ is the greatest common divisor $d(x)$:

$$d(x) = r_0(x)R_0(x) + r_1(x)R_1(x). \qquad (11.1\text{-}6)$$

Finally, let $r_0(x)$ be the polynomial on the left of (11.1-3) and let $r_1(x)$ be any polynomial $q(x)$ of degree $n - 1$. Since $r_0(x)$ is irreducible, the GCD $d(x)$ is a constant, and it can be taken equal to 1, because $R_0(x)$ and $R_1(x)$ in the above equation can be multiplied by any desired nonzero rational constant. Since $r_0(\alpha) = 0$, setting $x = \alpha$ in (11.1-6) gives

$$\frac{1}{q(\alpha)} = \frac{1}{r_1(\alpha)} = R_1(\alpha).$$

That enables us to get rid of the denominator mentioned above; hence (11.1-4) is established for any x in $\mathbb{F}(\alpha)$.

4. Show that a primitive element θ of the field $\mathbb{Q}(\sqrt{2}, \sqrt{3})$ is $\theta = \sqrt{2} + \sqrt{3}$; that is, show that $\sqrt{2}$ and $\sqrt{3}$ can both be expressed in terms of θ. Find an irreducible equation with rational coefficients satisfied by θ. Similar for $\mathbb{Q}(\sqrt{2}, \sqrt{3}, \sqrt{5})$.

5. It can be proved that if $\mathbb{F}(\alpha) = \mathbb{F}(\alpha')$, so that the extension can be obtained by adjoining either α or α', which are algebraic numbers over $\mathbb{F}$, then α and α' have the same degree as algebraic numbers, so that the degree of the extension is independent of which element we adjoin. The purpose of this exercise is to prove a more general result. If $\mathbb{K}$ is a finite algebraic extension of $\mathbb{F}$, then a *basis* of the extension is a set $\{\xi_1, \dots, \xi_n\}$ of numbers in $\mathbb{K}$ such that any number x in $\mathbb{K}$ can be uniquely expressed as a linear combination of $\xi_1, \dots, \xi_n$ with coefficients in $\mathbb{F}$. For example, according to (11.1-4), if $\mathbb{K} = \mathbb{F}(\alpha)$, then the set of numbers $1, \alpha, \alpha^2, \dots, \alpha^{n-1}$ is a basis. Prove that if $\{\xi_1, \dots, \xi_n\}$ and $\{\eta_1, \dots, \eta_k\}$ are two bases for $\mathbb{K}$ as an extension of $\mathbb{F}$, then $k = n$. *Hint:* If $k > n$, express the first n of the η_i as $\eta_i = \sum_{j=1}^{n} a_{ij}\xi_n$ $(i = 1, \dots, n)$. Since all of the η_i are linearly independent (that follows from the uniqueness in the definition), a sum $\sum_{i=1}^{n} y_i \eta_i = 0$ only if all the y_i are zero. Show then that in matrix-vector notation, $\vec{y}^T A = 0$ only if $\vec{y} = 0$, where A is the matrix of the a_{jk} in the above expression. From linear algebra it is known that then the matrix A is nonsingular, so that the ξ_i can be expressed in terms of the first n of the η_i; hence any x in $\mathbb{K}$ can be expressed in terms of the first n of the η_i. Therefore k cannot exceed n.

6. Show that if n is the degree of $\mathbb{F}(\alpha)$ over $\mathbb{F}$ and m is the degree of $\mathbb{F}(\alpha, \beta)$ over $\mathbb{F}(\alpha)$, the nm is the degree of $I\!F(\alpha, \beta)$ over $\mathbb{F}$. *Hint:* The quantities $\alpha^i \beta^j$ form a basis for $\mathbb{F}(\alpha, \beta)$ over $\mathbb{F}$.

11.2 Normal Sets of Points

Definition 11.2: A set of points in the hyperbolic plane is *normal* if the distance y between any two of them is such that e^y is in a quadratic-surd field and the angle α determined by any three of them is such that $\tan\alpha$ is in a quadratic-surd field.

We shall prove that when any additional point is constructed from a normal set by use of the straightedge and compass, the enlarged set continues to be normal. That will prove the first main result (Theorem 11.1 below).

As in the preceding section, r, θ denote polar coordinates relative to the origin Z and the line ℓ given at the beginning. If the set that constitutes the figure at any stage of its construction is normal, then, in particular, for any point r, θ of the set, e^r and $\tan\theta$ are in a quadratic-surd field. We shall show conversely that if that is true for each point of the set, the set is normal. To do that, we must show first that if r, θ and r', θ' represent any two points such that $e^r, e^{r'}, \tan\theta$, and $\tan\theta'$ are in such a field and if y is the distance between them, then e^y is in such a field. Then we must show that if α is the angle determined by any three such points, then $\tan\alpha$ is in such a field. Then, our main task will reduce to showing that whenever a new point with coordinates r, θ is added to a normal figure by use of the straightedge or compass, e^r and $\tan\theta$ are in such a field, and it will follow that the figure continues to be normal. We shall make repeated use of the following observations.

Observation 1. If one of the numbers $e^y, \sinh y, \cosh y, \tanh y$ is in a quadratic-surd field, then so are the others.

Proof: First, if e^y is in such a field, then so are the others, because they are rational functions of e^y. Second, if one of the hyperbolic functions is in such a field, then e^y can be obtained from it by solving a quadratic equation, and then the other functions are obtained from e^y by rational operations. □

Observation 2. If one of $\sin\alpha, \cos\alpha$, or $\tan\alpha$ is in a quadratic-surd field, then so are the others, unless $\cos\alpha$ is zero, in which case $\sin\alpha, \cos\alpha$, and $\cot\alpha$ are in such a field.

The proof is similar and is left to the reader.

Lemma 1: If, relative to some polar coordinate system, a set of points is such that for each point, e^r and $\tan\theta$ are in a quadratic-surd field, then the set is normal.

Proof. As noted above, we must prove first that if P and Q are any points of the set and $y = |PQ|$, then e^y is in such a field. Let r, θ be the coordinates of P and r', θ' those of Q. According to the generalized law of cosines,

Equation (6.3-5), we have

$$\cosh |PQ| = \cosh r \cosh r' - \sinh r \sinh r' \cos(\theta - \theta').$$

The quantities on the right are all in a quadratic-surd field, hence so is $e^{|PQ|}$. Second, it must be proved that if P, Q, R are any noncollinear points of the set referred to in the lemma and $\alpha = \angle PQR$, then $\tan \alpha$ is in a quadratic-surd field. Call $a = |QP|$, $b = |QR|$ and $c = |PR|$, as in Fig. 11.2a. We use the generalized law of cosines again:

$$\cosh c = \cosh a \cosh b - \sinh a \sinh b \cos \alpha,$$

from which we see, since $\sinh a$ and $\sinh b$ are $\neq 0$, that $\cos \alpha$ is in a quadratic-surd field. This completes the proof of the lemma. $\square$

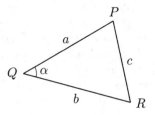

Fig. 11.2a.

We now consider the process of adding points to a normal figure by straightedge and compass. A straightedge can draw a line and a compass can draw a circle; any new point that can be added is either the intersection of two lines, the intersection of a line and a circle, or the intersection of two circles. As noted above, we can assume that after an initial construction as in Fig. 11.1a no lengths are used (i.e., as radii of circles) except ones already present as distances in the point set. The general equation of a line in polar coordinates r, θ is (6.4-1), which we write as

$$\tanh r \cos(\theta - \alpha) = B, \tag{11.2-1}$$

where B and α are constants such that either $0 < B < 1$ and $|\theta - \alpha|$ is less than $\arccos B$, or $B = 0$ and $\theta = \alpha + \pi/2 \pmod{\pi}$; in the latter case, the line passes through the origin. We assume that the line was determined by two given points, with polar coordinates r_1, θ_1 and r_2, θ_2. Since the line passes through those points, we have

$$\tanh r_1(\cos \theta_1 \cos \alpha + \sin \theta_1 \sin \alpha) = B,$$
$$\tanh r_2(\cos \theta_2 \cos \alpha + \sin \theta_2 \sin \alpha) = B. \tag{11.2-2}$$

We can equate these two expressions and solve for $\tan \alpha$ (or $\cot \alpha$ in case $\tan \alpha = \infty$). We assume that the given points were in a normal figure, so that $\tanh r_1, \tanh r_2$ and the trigonometric functions of θ_1 and θ_2 are in a quadratic-surd field. Hence, $\tan \alpha$ (or $\cot \alpha$) is in such a field, and it then follows from either of the equations above that B is also in such a field.

We now consider two such lines, given by constants B, B', α, α'. We assume that they intersect at a point P and we seek the coordinates r, θ of that point. We have

$$\tanh r(\cos \theta \cos \alpha + \sin \theta \sin \alpha) = B,$$
$$\tanh r(\cos \theta \cos \alpha' + \sin \theta \sin \alpha') = B'. \tag{11.2-3}$$

By equating the two resulting expressions for $\coth r$, we get

$$\frac{\cos \theta \cos \alpha + \sin \theta \sin \alpha}{B} = \frac{\cos \theta \cos \alpha' + \sin \theta \sin \alpha'}{B'}, \tag{11.2-4}$$

and it follows that $\tan \theta$ is in a quadratic-surd field, and then it follows from either of the above equations that $\tanh r$ is also in such a field. Therefore, if the point P is added to the set, the set continues to be normal.

Before considering intersections involving circles, we note that the normality of a figure is invariant under isometries in the following sense. Let P and Q be any two points of the figure; we can take P as a new origin and the ray $\overrightarrow{PQ}$ as the new axis of $\theta = 0$. Then, for any point of the figure, $\tanh r$ and $\tan \theta$ continue to be in a quadratic-surd field. Then, if P is taken as the center of a circle, the equation of the circle in the new polar coordinates is $r = a$, where a is the distance between two of the points of the figure, so that $\tanh a$ is in such a field. For the intersection of a line and a circle, we then have

$$\tanh r = \frac{B}{\cos(\theta - \alpha)}$$

and $r = a$, from which we see that the quantities $\tanh r$ and $\tan \theta$ of the intersection are in such a field.

Now consider the intersection of two circles with centers at P and Q and radii a and b. The point X of intersection is given by

$$|PX| = a, \qquad |QX| = b.$$

We denote the distance between centers by $c = |PQ|$ as in Fig. 11.2b.

In terms of polar coordinates r, θ determined by the points P, Q as above, the coordinates of the point X are $r = a$, $\theta = \angle XPQ$. If the circles intersect, as assumed, then $a + b \geqq c$. If $a + b = c$, the circles are tangent, and $\theta = 0$. If $a + b > c$, we have

$$\cosh b = \cosh a \cosh c - \sinh a \sinh c \cos \theta,$$

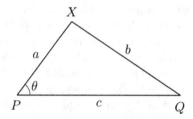

Fig. 11.2b.

and we see that $\cos\theta$ is in a quadratic-surd field. In either case $e^r = e^a$; hence e^r and $\cos\theta$ are in a quadratic-surd field. We have proved the following.

Theorem 11.1: Suppose that the points and lines given prior to a construction constitute a normal figure. Then, any figure obtained therefrom by construction with straightedge and compass is normal. That is, if y is any length in the resulting figure and α is any angle, then e^y and $\tan\alpha$ are in a quadratic-surd field.

We note the curious fact that a segment of unit length cannot be constructed, because e is not even an algebraic number. The converse of Theorem 11.1 is Theorem 11.4 in Section 11.6 below.

11.3 Segment Trisection

We now show that in hyperbolic plane geometry, the trisection of a straight-line segment is not always possible. We recall that the trisection is done in Euclidean geometry by use of parallel lines, as indicated in Fig. 11.3a. (Division into n parts is similar.)

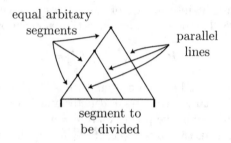

Fig. 11.3a.

In hyperbolic geometry, one does not have unique parallels, and they don't do what one expects, anyway. It was shown above that the length $y_0 = \ln(\sqrt{2}+1)$ can be constructed, and we shall show that $(1/3)\, y_0$ cannot, by showing that the number $\exp(1/3)\, y_0)$, that is, $(\sqrt{2}+1)^{1/3}$, is not in a quadratic-surd field. It was shown above that $\sinh y_0 = 1$. If $\exp((1/3)\, y_0)$ were in such a field, then $\sinh(1/3)\, y_0$ would be also, but we can compute $\sinh y_0$ from $\sinh(1/3)\, y_0 \overset{\text{def}}{=} x$ by the formula

$$1 = \sinh y_0 = 4x^3 + 3x.$$

This equation is irreducible. If $4x^3 + 3x - 1$ could be factored over the rational numbers, one of the factors would have to be $4x \pm 1$, $2x \pm 1$, or $x \pm 1$, but none of these divides $4x^3 + 3x - 1$. Therefore, x is an algebraic number of degree 3 over the rational numbers, while the degree of any element of a quadratic-surd field is a power of 2. Since 3 does not divide any power of 2, the number $x = \sinh(1/3)\, y_0$ is not in a quadratic-surd field; hence $(1/3)\, y_0$ is not constructible.

This nontrisectibility is discussed in various places, for example in *The Foundations of Geometry and the Non-Euclidean Plane*, by George Martin, p. 483, (Springer Verlag, Berlin, 1975), and in *Euclidean and Non-Euclidean Geometry*, by Marvin J. Greenberg, p. 176, (W. H. Freeman and Company, 1974). We have not been able to learn who discovered it first.

Exercise

1.   Show that the problem of angle trisectibility is the same as in Euclidean geometry. *Hints*: The two geometries have the same requirement for the trigonometric functions of angles. Try to trisect 60°. Why is 90° trisectible?

11.4 Construction of the Angle of Parallelism

The constructions described in Section 11.1 had to do with the angle of parallelism $\alpha = \Pi(y)$; it was shown how to construct the length y when the angle α is given. The converse problem of constructing α when y is given was solved by Bolyai. A simple version of Bolyai's construction is this. Draw the Lambert quadrilateral shown in Fig. 11.4a, with right angles at A, B, C and with $|AB|$ and $|BC|$ both equal to the given length y.

   The other two sides have a greater length y'. With the compass set at radius y', draw a circle about point B. The intersection of that circle with side CD is called X, and it is claimed that the angle $\alpha = \angle ABX$ is equal to $\Pi(y)$. To show that, denote by d the length of the diagonal BD. By symmetry, the angle $\angle ABD$ is half of $\angle ABC$, hence is equal to 45°. By the triangle functions, we have

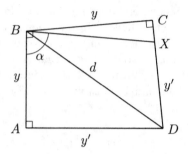

Fig. 11.4a.

$$\tanh y = (\tanh d)(1/\sqrt{2}),$$
$$\cosh d = \cosh y \cosh y',$$
$$\tanh y = \tanh y' \cos(90° - \alpha).$$

If we eliminate y' and d from these equations (which takes a fair amount of calculation), we find that $\tanh y = \cos \alpha$, which is one form of the equation for the angle of parallelism, namely, $\tanh y = \cos \Pi(y)$.

We note a subtle difference between this problem and the construction problems considered in the preceding section. The length y can be given in advance and does not itself have to be constructible in the sense of that section. Similarly, in the construction of y from α as described in Section 11.1, the angle α can be given in advance and does not have to be constructible.

Since the ray $\overrightarrow{BX}$ is asymptotic to the ray $\overrightarrow{AD}$, we have constructed the line through the finite point B and the ideal point at the end of $\overrightarrow{AD}$, without assuming in advance that a straightedge can connect those points.

11.5 Squaring the Circle

In Euclidean geometry a circle of radius r has the same area as a square of sidelength c if $\pi r^2 = c^2$. The impossibility of squaring the circle, that is, of constructing c when r is given, comes from the fact that the ratio $c/r = \sqrt{\pi}$ is not in a quadratic-surd field (and is not even algebraic). That is independent of the size of the given circle, because the ratio c/r is a constant.

In hyperbolic geometry, for equal areas, the ratio c/r is not a constant. As $c \to \infty$ the area of a square,[4] which is equal to its angular defect,

---

[4] By a "square" is meant a quadrilateral with all sides of the same length and all angles congruent (hence less than 90°).

converges to the value 2π, while the area of the circle is equal to $2\pi(\cosh r - 1)$. Hence, for equal areas, r approaches a finite value such that $\cosh r = 2$, as $c \to \infty$, hence $c/r \to \infty$. Whenever c/r is equal to a positive integer, c can be constructed by addition of segments of length r.

In the limit, for a circle of radius such that $\cosh r = 2$, hence $r = \ln(2 + \sqrt{3})$, the square that has the same area is a degenerate square with its vertices at ideal points at infinity. According to Section 11.1, the distance from the center of the square to each side is then equal to $\ln(\sqrt{2} + 1)$ (see Fig. 11.1a). In this case, both the square and the circle can be constructed in the sense of Section 11.2.

Exercises

1. Write the area of a square of side c as $2\pi - 8\alpha$, where α is one-half the angle at one of the vertices of the square as in Fig. 11.5a. Derive an equation for α (rather, for $\sin \alpha$) by drawing the diagonals of the square and calling $d =$ one-half the length of a diagonal and then applying the triangle equations to each of the isosceles right triangles with right angles at the center of the square and eliminating d from those equations.

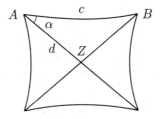

Fig. 11.5a.

2. Show how to construct a circle of radius $r = \ln(2 + \sqrt{3})$, that is, how to construct a segment of that length.
3. Discuss the problem of constructing a square when its side length c is given, or when its semidiagonal d is given, or when the distance from its center to a side is given.

11.6 Constructibility of All Points Associated with Quadratic-Surd Fields

Definition 11.3: Let $\mathbb{F}$ be a real field. A point in the hyperbolic plane with polar coordinates r, θ is *associated with* $\mathbb{F}$ if $\tanh r \cos \theta$ and $\tanh r \sin \theta$ are both in $\mathbb{F}$.

Definition 11.4: A line is *associated with* $\mathbb{F}$ if it is determined by (i.e., contains) two points associated with $\mathbb{F}$.

Theorem 11.2: If two lines are associated with a field $\mathbb{F}$, then their point of intersection is associated with $\mathbb{F}$.

[In other words, we cannot get out of the set of all points associated with $\mathbb{F}$ by constructions using the straightedge alone.]

Recall that the polar coordinate equation of a line was given as (11.2-1), namely,

$$\tanh r \cos(\theta - \alpha) = B. \tag{11.6-1}$$

In the proof of Theorem 11.2 we need the following lemma, whose converse is Lemma 7 below.

Lemma 2: If the line given by the above equation is associated with $\mathbb{F}$, but does not pass through the origin, then $(\sin \alpha)/B$ and $(\cos \alpha)/B$ are in $\mathbb{F}$. If the line passes through the origin, then $\tan \alpha$ is either in $\mathbb{F}$ or is infinite.

Proof: Let r_1, θ_1 and r_2, θ_2 be the points associated with $\mathbb{F}$ that lie on the line according to Definition 11.2. Then, equations (11.2-2) hold. If the line does not pass through the origin, B is not 0, and we can divide through by it. We have

$$\begin{aligned}
\tanh r_1 \left(\cos \theta_1 \frac{\cos \alpha}{B} + \sin \theta_1 \frac{\sin \alpha}{B} \right) &= 1, \\
\tanh r_2 \left(\cos \theta_2 \frac{\cos \alpha}{B} + \sin \theta_2 \frac{\sin \alpha}{B} \right) &= 1.
\end{aligned} \tag{11.6-2}$$

This is a linear system with coefficients in $\mathbb{F}$ for $(\cos \alpha)/B$ and $(\sin \alpha)/B$ as unknowns. Its determinant is $\tanh r_1 \tanh r_2 \sin(\theta_2 - \theta_1)$, which is $\neq 0$ because r_1 and r_2 are $\neq 0$ and if $\theta_2 - \theta_1$ were $\equiv 0 \pmod{\pi}$, the line would pass through the origin. Hence, we can solve for $(\cos \alpha)/B$ and $(\sin \alpha)/B$ as elements of $\mathbb{F}$, as required. If the line passes through the origin, then the first of the above equations can be replaced by $\tanh r_1 (\cos \theta_1 \cos \alpha + \sin \theta_1 \tan \alpha) = 0$, hence

$$\tan \alpha = -\frac{\tanh r_1 \cos \theta_1}{\tanh r_1 \sin \theta_1},$$

which is either in $\mathbb{F}$ or infinite.   □

Proof of Theorem 11.2: Let the first line be given by constants α and B and the second by constants α' and B'. Their intersection is the point r, θ given by

$$\tanh r(\cos\theta\cos\alpha + \sin\theta\sin\alpha) = B,$$
$$\tanh r(\cos\theta\cos\alpha' + \sin\theta\sin\alpha') = B'. \qquad (11.6\text{-}3)$$

First, if neither of the lines passes through the origin, we can divide the first equation through by B and the second by B'. The result is a linear system for the unknowns $\tanh r\cos\theta$ and $\tanh r\sin\theta$. The solution exists, according to the hypotheses of the theorem, and is unique, according to the axioms. According to Lemma 2, the coefficients of the system are in $\mathbb{F}$, hence those unknowns are in $\mathbb{F}$, and the point of intersection is associated with $\mathbb{F}$ according to Definition 11.2. If the first line passes through the origin, then the first equation of the system can be replaced either by $\tanh r(\cos\theta + \sin\theta\tan\alpha) = 0$ or by $\tanh r(\cos\theta\cot\alpha + \sin\theta) = 0$, so we have again a linear system for the same unknowns, with coefficients in $\mathbb{F}$. Lastly, if both lines pass through the origin, then the intersection is the origin. $\qquad \square$

Remark: From now on, we shall allow Definition 1 to be interpreted as including ideal points at infinity (where $\tanh r = 1$), if $\sin\theta$ and $\cos\theta$ are in $\mathbb{F}$. Then Lemma 1 continues to be true and so does Theorem 11.2.

Note: The quantities $X = \tanh r\cos\theta$ and $Y = \tanh r\sin\theta$ are the co-ordinates of the Beltrami–Klein model (Section 7.8) in which straight-line segments in the model represent straight-line segments in the hyperbolic plane. Therefore, any construction with straightedge alone in the model represents a construction with straightedge in the plane. Below, we shall consider sets of points in the model that form a square lattice. The corresponding set of points in the plane do not constitute such a lattice; although points that are in a line represent points in a line, the squares in the model represent various mostly nonsquare quadrilaterals in the plane.

Lemma 3: The four points where $X = \pm\frac{1}{2}$ and $Y = \pm\frac{1}{2}$ can be constructed.

Proof: The construction of the point where $X = Y = \frac{1}{2}$ was discussed in Section 11.1. We drew the line m perpendicular to ℓ at Z; then we connected the ideal points at the ends of ℓ and m by straightedge to give a line n. We then dropped a perpendicular from Z to n. We found that the foot of that perpendicular on n has polar coordinates r, θ such that $\sinh r = 1$ and $\theta = 45°$. Therefore $\tanh r = 1/\sqrt{2}$ and the sine and cosine of θ are also $1/\sqrt{2}$, so that $X = Y = \frac{1}{2}$. The construction of the other three points is similar.

Note: Those constructions use the compass and an arbitrary length a, but the result is independent of a, and after the construction is completed, we abandon the value of a. From now on, we use the straightedge alone for points associated with $\mathbb{Q}$.

Lemma 4: The points for which $X = 2^{-n}p$ and $Y = 2^{-n}q$, where n, p, q are integers and $p^2 + q^2 < 4^n$ can be constructed.

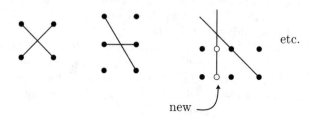

Fig. 11.6a.

Proof: We note first that the point with $X = \frac{1}{2}$ and $Y = 0$ can be constructed as the intersection of the line joining the points $((1/2), (1/2))$ and $((1/2), -(1/2))$ with the given line ℓ. The other three points where one of X, Y is $\pm(1/2)$ and the other is zero, can be similarly constructed. Since the origin was given and needs no construction, we now have the part of a square lattice with edge length $1/2$ that lies inside the unit circle. We now proceed by induction: Assume that a lattice with edge length 2^{-n} can be constructed. We then construct the points of a lattice with edge length 2^{-n-1} by the intersection of lines such as indicated in Fig. 11.6a. The open circles in the third drawing are points of the new lattice that have been constructed previously. Consideration of special cases shows that in this way all points of the new lattice that lie inside the unit circle $X^2 + Y^2 < 1$ can be constructed. Then, induction on n completes the proof of the lemma.   □

Theorem 11.3: All points associated with the field $\mathbb{Q}$ can be constructed.

Proof: Let ξ, η be rational numbers with $\xi^2 + \eta^2 < 1$. We show how to construct two lines that intersect at the point with Beltrami–Klein coordinates $X, Y = \xi, \eta$. Consider a square lattice with edgelength 2^{-n} small enough that the given point lies within one of the lattice squares in that part of the lattice inside the unit circle. Let ξ_0, η_0 be a lattice point near ξ, η (e.g., one of the corners of the square just referred to), and let ℓ be the line through the points ξ, η and ξ_0, η_0. We shall find another lattice point ξ_1, η_1 that lies on that line. We write

$$\xi = 2^{-n}f, \qquad \eta = 2^{-n}g,$$

where f and g are rational numbers, and

$$\xi_0 = 2^{-n}p, \qquad\qquad \eta_0 = 2^{-n}q,$$
$$\xi_1 = 2^{-n}(p + \Delta p), \quad \eta_1 = 2^{-n}(q + \Delta q),$$

where $p, q, \Delta p, \Delta q$ are integers. The slope of the line ℓ (i.e., its slope in the model) is $(\eta - \eta_0)/(\xi - \xi_0)$. If $\xi - \xi_0$ is zero, the line is vertical and ξ_1, η_1 can be taken as the lattice point immediately above or below ξ_0, η_0. Otherwise, for the point ξ_1, η_1 to lie on the line ℓ, we must have

$$\frac{\eta - \eta_0}{\xi - \xi_0} = \frac{\Delta q}{\Delta p};$$

since the left member is a rational number, we take the right member to be the fraction that gives that number in lowest terms. We now face the difficulty that the point ξ_1, η_1 may not lie in the unit circle. We remedy that possibility as follows: We increase n, keeping ξ, η, ξ_0, η_0 fixed. Then, f, g, p, q will increase, but we keep Δp and Δq fixed as given by the equation above. Then, the distance between the points ξ_0, η_0 and ξ_1, η_1 is $2^{-n}[\Delta p^2 + \Delta q^2]^{1/2}$. If n is large enough, this will be less than the distance from ξ_0, η_0 to the unit circle, and the new point lies inside, so that the line ℓ can be constructed, when the square lattice is given. We then find another line ℓ' similarly, and the point ξ, η is the intersection of the two lines. That proves Theorem 11.3.   □

We now consider the problem of constructing extensions of a point set associated with a given field $\mathbb{F}$.

Lemma 5: If points P and Q are associated with a field $\mathbb{F}$, then $(\tanh|PQ|)^2$ is in $\mathbb{F}$.

Proof: Since P and Q are associated with $\mathbb{F}$, their Beltrami–Klein coordinates

$$X_i = \tanh r_i \cos \theta_i \qquad Y_i = \tanh r_i \sin \theta_i \qquad (i = 1, 2)$$

are in $\mathbb{F}$. We shall show that $(\tanh|PQ|)^2$ can be expressed in terms of them. As in the proof of Lemma 1 in Section 11.2, we use the generalized law of cosines (6.3-5)

$$\cosh|PQ| = \cosh r_1 \cosh r_2 - \sinh r_1 \sinh r_2 \cos(\theta_2 - \theta_1).$$

We square this and subtract 1, to get $\sinh^2|PQ|$; we write the "1" in a complicated way, knowing that, for any x, $\cosh^2 x - \sinh^2 x = 1$:

$$\sinh^2|PQ| = [\cosh r_1 \cosh r_2 - \sinh r_1 \sinh r_2 \cos(\theta_2 - \theta_1)]^2$$
$$- (\cosh^2 r_1 - \sinh^2 r_1)(\cosh^2 r_2 - \sinh^2 r_2).$$

We divide through by $\cosh^2|PQ|$, which is just the first term on the right of the above equation, to give $\tanh^2|PQ| = 1-$ a fraction; in the fraction, we divide numerator and denominator by $\cosh^2 r_1 \cosh^2 r_2$, and we have

$$\tanh^2 |PQ| = 1 - \frac{(1 - \tanh^2 r_2)(1 - \tanh^2 r_1)}{[1 - \tanh r_1 \tanh r_2 \cos(\theta_2 - \theta_1)]^2},$$

$$= 1 - \frac{(1 - X_2^2 - Y_2^2)(1 - X_1^2 - Y_1^2)}{[1 - X_1 X_2 - Y_1 Y_2]^2},$$

which is in $\mathbb{F}$. □

We shall need the converse of Lemma 2, and to prove it, we need an intermediate step:

Lemma 6: The angles θ such that $\sin\theta$ and $\cos\theta$ are rational are dense in $[0, 2\pi]$.

Proof: Let m and n be integers, and write

$$\sin\theta = \frac{2mn}{m^2 + n^2}, \qquad \cos\theta = \frac{m^2 - n^2}{m^2 + n^2}.$$

It is clear that the resulting values of $\sin\theta$ and $\cos\theta$ are dense in $[0, 1]$ and satisfy $\sin^2\theta + \cos^2\theta = 1$. □

Lemma 7: If a line with equation (11.2-1) is such that $(\sin\alpha)/B$ and $(\cos\alpha)/B$ are in $\mathbb{F}$, then it is associated with $\mathbb{F}$ (i.e., contains two points that are associated with $\mathbb{F}$).

Proof: Let θ_1 and θ_2 be angles such that their sines and cosines are rational and are in the open interval from $\alpha - \arccos B$ to $\alpha + \arccos B$. Then points r_1, θ_1 and r_2, θ_2 are on the line if equations (11.2-2) are satisfied. On dividing (11.2-2) through by B, we see that the quantities in the parentheses in that equation are in $\mathbb{F}$ and are greater than 1, so that the equations can be solved for values of $\tanh r_1$ and $\tanh r_2$ in $\mathbb{F}$. Since the sines and cosines of θ_1 and θ_2 are in $\mathbb{F}$, the Beltrami–Klein coordinates of the points are in $\mathbb{F}$. □

We now consider intersections of lines and circles. We recall that in the Euclidean case, if we construct a circle of radius r and intersect it with a line at a distance x ($< r$) from the center of the circle, where r and x are in a field $\mathbb{F}$, then the length of the chord of the circle so formed can be the square root of any positive number in $\mathbb{F}$. A similar thing is true in the hyperbolic case.

In terms of polar coordinates r, θ, the equation of a line perpendicular to the horizontal axis at the point where $r = b$ and $\theta = 0$ is

$$\tanh r \cos\theta = B = \tanh b.$$

The intersection of that line with a circle centered at the origin and of radius a is the point a, θ, where θ is the solution of the above equation with $r = a$. Then,

$$\tanh a \sin \theta = \sqrt{(\tanh a)^2 - (\tanh b)^2} \overset{\text{def}}{=} \sqrt{w}. \tag{11.6-4}$$

The point P of intersection has Beltrami–Klein coordinates X, Y, where $X = \tanh a \cos \theta$ and $Y = \tanh a \sin \theta$. We recall that although the model is not conformal, orthogonal lines in the model represent orthogonal lines in the plane if (and only if) one of the lines passes through the origin. Therefore, we can construct the point with Beltrami–Klein coordinates $0, Y$ by dropping a perpendicular from the point P of intersection to the line $\theta = \pi/2$, where $\tanh d = \sqrt{w}$. If $\tanh a$ and $\tanh b$ are in a field $\mathbb{F}$, then w is in $\mathbb{F}$. Conversely, if w is in $\mathbb{F}$ and $0 < w < 1$, $\tanh a$ and $\tanh b$ can be so chosen in $\mathbb{F}$ that the equation

$$(\tanh a)^2 - (\tanh b)^2 = w$$

holds. For example, one can set $\tanh a = (1 + w)/2$ and $\tanh b = (1 - w)/2$. We have proved the following:

Lemma 8: If all points associated with a field $\mathbb{F}$ can be constructed, then one can construct any length d such that $\tanh d = \sqrt{w}$, where w is an element of $\mathbb{F}$ such that $0 < w < 1$.

That is the first step in proving that all points associated with the field $\mathbb{F}(\sqrt{w})$ can be constructed. We show next that certain distances d such that $\tanh d = c_1 + c_2 \sqrt{w}$ can be constructed, with c_1 and c_2 in $\mathbb{F}$. That is done by intersecting an arbitrary line associated with $\mathbb{F}$ with a circle of radius R about the origin. For constructing the circle, it is not necessary that $\tanh R$ be in $\mathbb{F}$; according to the lemma above, it is only necessary that $(\tanh R)^2$ be in $\mathbb{F}$. The method is this. Let c_1, c_2, w be elements of $\mathbb{F}$ such that $(c_1^2 + c_2^2)(1 + w)$ is less than 1. Define B, α, and R by the equations

$$B = \sqrt{c_1^2 + c_2^2}, \tag{11.6-5}$$
$$\sin \alpha = c_1/B, \quad \cos \alpha = c_2/B \tag{11.6-6}$$
$$\tanh^2 R = B^2(1 + w). \tag{11.6-7}$$

Then $(\cos \alpha)/B$ and $(\sin \alpha)/B$ are in $\mathbb{F}$, so that the line with constants α, B is associated with $\mathbb{F}$. The intersection of that line with the circle $r = R$ is the point R, θ, where θ is determined from the equation

$$\tanh R(\cos \theta \cos \alpha + \sin \theta \sin \alpha) = B.$$

This gives a quadratic equation for $\sin \theta$, namely,

$$\tanh^2 R \sin^2 \theta - 2B \tanh R \sin \alpha \sin \theta + B^2 - \tanh^2 R \cos^2 \alpha = 0$$

and a similar equation for $\cos \theta$. From them, we find for the Beltrami–Klein coordinates of the intersection

$$Y = \tanh R \sin\theta = c_1 \pm c_2\sqrt{w},$$
$$X = \tanh R \cos\theta = c_2 \mp c_1\sqrt{w}. \tag{11.6-8}$$

One intersection is given by the upper signs, the other by the lower.

We recall that one can drop a perpendicular from a given point to a given line in the hyperbolic plane and that although the model is not conformal two orthogonal line segments in the model represent orthogonal line segments in the plane if (and only if) one of them passes through the origin. Therefore, we can project either of the above points onto the X and Y axes, and we have the following result.

Lemma 9: If all points associated with $\mathbb{F}$ can be constructed and if w, c_1, and c_2 are numbers in $\mathbb{F}$ such that $0 < w < 1$ and $c_1^2 + c_2^2 < 1/(1+w)$, then the length d such that $\tanh d = c_1 + c_2\sqrt{w}$ can be constructed.

To get rid of the restriction on $c_1^2 + c_2^2$, we now let $a_1 + a_2\sqrt{w}$, with a_1 and a_2 in $\mathbb{F}$, be any number in $\mathbb{F}(\sqrt{w})$ such that $|a_1 + a_2\sqrt{w}| < 1$. We define $c_1 = a_1/n$, $c_2 = a_2/n$, where n is a positive integer large enough so that c_1 and c_2 satisfy the restriction in the lemma and so that the number $h = c_1 + c_2\sqrt{w}$ also satisfies $|h| < \frac{1}{2}$. We construct the eight points with Beltrami–Klein coordinates $(X, Y) = (\pm h, \pm h)$, $(0, \pm h)$, and $(\pm h, 0)$. These points, together with the origin, constitute the nucleus of a square grid of points in the model with side length h; they are shown as solid points in Fig. 11.6b. By means of the straightedge, the points indicated by open circles can be constructed in the order indicated by the numbering. This procedure can be continued until all of the lattice points inside the unit circle have been constructed. Among the points is one with $X = nh = a_1 + a_2\sqrt{w}$, showing that the length d such that $\tanh d = a_1 + a_2\sqrt{w}$ can be constructed. The conclusion is this.

Lemma 10: If all points associated with a field $\mathbb{F}$ can be constructed, and if w is an element of $\mathbb{F}$ such that $0 < w < 1$, then all points associated with $\mathbb{F}(\sqrt{w})$ can be constructed.

We note that the restriction $w < 1$ is not essential, because a field $\mathbb{F}(\sqrt{w})$ is identical with the field $\mathbb{F}(\sqrt{w}/q)$, where q is any integer.

We now consider general quadratic-surd fields, as defined inductively by Equations (11.1-5). As just noted, we can assume, without loss of generality, that each of the w_i is in the interval $(0, 1)$. Then by an induction on i, in which the first step is Theorem 11.3 (all points associated with $\mathbb{Q}$ can be constructed) and the inductive step is Lemma 10, we arrive at the final result:

Theorem 11.4: All points in the hyperbolic plane associated with a quadratic-surd field can be constructed by straightedge and compass.

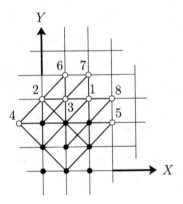

Fig. 11.6b.

This was stated by Mordukhoy–Boltovski in the form that any length d such that $\sinh d$ is in a quadratic-surd field can be constructed.

Exercise

1. In Lemma 4, how many of the points of the lattice for $p = 2$ lie inside the unit circle? Show how to construct those points by straightedge from the nine points of the lattice for $p = 1$ that lie inside the unit circle.

11.7 Construction of Regular Polygons

It is recalled that Gauss proved that a regular polygon of an odd number N of sides can be constructed by straightedge and compass in Euclidean geometry if and only if N is a Fermat prime or a product of distinct Fermat primes. See Oystein Ore, *Invitation to Number Theory*, New Mathematical Library #20, Random House, New York, 1967. A *Fermat number* is a number of the form

$$F_n = 2^{2^n} + 1,$$

where n is a nonnegative integer. The known primes among them are F_0, F_1, F_2, F_3, and F_4. F_5 is a composite. As of 1967, it was apparently conjectured by some mathematicians that F_n is composite for all $n \geq 5$.

In particular, it follows from Gauss' theorem that a regular N-gon is constructible for $N = 3, 5, 15, 17$, but not for $N = 7, 9, 11, 13$. For a proof of Gauss' theorem, see *Modern Algebra*, by B. L. van der Waerden, Section 59, Frederick Ungar, 1949.

Clearly, a regular polygon of $2^k N$ sides, where N is odd, can be constructed if and only if one with N sides can be constructed. (The "if" part comes from bisecting angles, the "only if" part by ignoring some of the vertices.)

In the Euclidean case, the polygons are usually thought of as inscribed in the unit circle. In the hyperbolic case, the radius of the circle enters as a new parameter, because a regular N-gon inscribed in a circle of one radius is not similar to one (with the same N) inscribed in a circle of a larger radius. The latter has smaller internal angles, for given N. We shall assume that the radius a is a constructible length, so that e^a is in a quadratic-surd field. With that condition, we shall see, regular N-gons are constructible for the same values of N as in the Euclidean case, namely, the ones that follow from Gauss' theorem.

In the Euclidean case, if the circle has unit radius and if c is a side of the N-gon, and γ denotes the angle $2\pi/N$, we have $c/2 = \sin(\gamma/2)$. Hence, then, c is in a quadratic-surd field if and only if $\sin(\gamma/2)$ is; hence, Gauss' theorem is equivalent to saying that $\sin(\pi/N)$ is in a quadratic-surd field if and only if N is of the form 2^k times a Fermat prime or a product of distinct Fermat primes. Now, it is an immediate corollary of Theorem 11.4 in the preceding section that an angle α is constructible in the hyperbolic plane if and only if $\tan\alpha$ (or equivalently $\sin\alpha$ or $\cos\alpha$) is in a quadratic-surd field. In the hyperbolic case, we have ·

$$\sinh\frac{c}{2} = \sinh a \sin\frac{\gamma}{2},$$

where a is the radius of the circle and c is the side of the N-gon. Therefore, if a is constructible, c is constructible for precisely the same values of N as in Gauss' theorem.

11.8 The Horocompass Gives Nothing New

The words "straightedge" and "compass" indicate merely that lines and circles are used in the abstract construction process, and one may wonder whether a larger class of curves could be used, thereby strengthening the final theorem, which says that those curves lead precisely to those figures that can be described in terms of quantities in quadratic-surd number fields. It is easy to see that one cannot enlarge the class to include all conic sections, because one can find a parabola and a hyperbola, both satisfying equations with rational coefficients, whose intersection has coordinates x, y that satisfy irreducible cubic equations (Exercise 1 below). The question arises, what properties do lines and circles have, geometrically, that distinguish them from all other curves? One possible answer, in the Euclidean case, is that they are exactly the self-similar curves.

Definition 11.5: A curve is *self-similar* if it is invariant (i.e., is mapped onto itself) under a nontrivial direct isometry of the geometry (hence is invariant under a one-parameter subgroup of the isometry group).

In the hyperbolic plane, the self-similar curves are lines, circles, horocycles, and equidistants; see Section 8.2. According to the concluding section of the excellent book *Non-Euclidean Geometry* by H. S. M. Coxeter, University of Toronto Press, 1961, the use of horocycles and equidistants in the constructions was investigated by Nestorowitsch, who proved in 1949 that their use does not lead to any figures that cannot be constructed by straightedge and compass alone.

In the other direction, we recall that it was proved by Mascheroni in the eighteenth century that any set of points that can be constructed by straightedge and compass can be constructed by compass alone. For similar questions in the hyperbolic case, we refer again to the book by Coxeter.

A question that arises concerns the coordinate systems that can be used. In the Euclidean case, it is natural to use Cartesian coordinates; in fact, Cartesian coordinates based on the initially given origin and additional point. In the hyperbolic case, we used the Beltrami–Klein coordinates $\tanh r \cos\theta$ and $\tanh r \sin\theta$. If we were to use those coordinates in the Euclidean case or Cartesian coordinates $x = r\cos\theta$, $y = r\sin\theta$ in the hyperbolic case, we should encounter coordinate values not in quadratic-surd fields, and not even algebraic. Surely one must require the coordinates to be such that the self-similar curves satisfy algebraic equations. For the method of proof used in Section 11.6 above, it was even essential also that lines satisfy linear equations.

As noted earlier, there are also construction problems in which one is given something other than the minimal figure mentioned above (a line and a point on it in the hyperbolic case, two points in the Euclidean case). Examples are: (1) given two lines, find a line asymptotic to one and perpendicular to the other; (2) given three points, determine by construction whether they lie on a line, on an equidistant, on a horocycle, or on a circle; in the second case, construct the line to which the curve is equidistant; in the fourth case, construct the center of the circle. (3) given two ultraparallel lines (i.e., nonintersecting and not asymptotic), construct their common perpendicular.

Exercises

1. Show that the coordinates of the point of intersection of the curves

    $$xy + 2x + 2 = 0, \qquad y = x^2,$$

    satisfy irreducible cubic equations. *Hint*: Recall Gauss' theorem that says that if a polynomial with integer coefficients cannot be factored over the integers, it cannot be factored over the rationals.

2. Let r, θ be polar coordinates in the hyperbolic plane, and let the line given by

    $$\tanh r \cos(\theta - \alpha) = B$$

    be associated with the rational field $\mathbb{Q}$. Show that the Beltrami–Klein coordinates of its intersection with the horocycle given by

$$\tanh \frac{r}{2} = \cos\theta$$

are in a quadratic-surd field.

3.   Given two lines that are ultraparallel. Show how to find their common perpendicular by straightedge and compass, where it is assumed that the straightedge can join ideal points. *Hint*: Let P and P' be the ideal points at the ends of the first given line, and Q and Q' the ideal points at the end of the other. Construct the lines joining P and Q' and joining P' and Q.

4.   Suppose the points A, B, C lie on an equidistant to a line ℓ. Show how to construct the line ℓ when the points A, B, C are given. *Hint*: The perpendicular bisectors of the segments AB and BC are ultraparallel. See Exercises 1 and 2 in Section 3.10.

11.9 The Finite Straightedge

Our assumption that a straightedge can draw the line between two ideal points at infinity is conceptually reasonable, because it corresponds to the fact that in the hyperbolic plane two ideal points determine a unique line. It also greatly simplifies many constructions, such as finding the common perpendicular to ultraparallel lines and finding a line asymptotic in one direction to one side of a given angle and asymptotic in the other direction to the other side of the angle (see Fig. 3.4c).

However, that assumption is not necessary. All constructions are possible with compass and a finite straightedge (one that can draw the line through any two finite points). The constructions are generally more complicated than the ones we have described; we give brief descriptions in the exercises below. For more detail we refer again to the book by Coxeter.

In Section 11.4 we described Bolyai's construction of a line through a given point asymptotic to a given line. It uses only the finite straightedge. Next, a method was given by Hilbert in 1913 for constructing the common perpendicular to a pair of ultraparallel lines. See Exercise 1 below. (As indicated in Exercise 3 in the preceding section, that construction is very easy with the infinite straightedge.) Finally, a method was given, also by Hilbert in 1913, for solving the following problem. Given two nonasymptotic lines (i.e., lines either intersecting or ultraparallel), construct the line m that is asymptotic to one of the given lines in one direction on m and asymptotic to the other in the other direction. It uses both of the foregoing constructions. See Exercise 2.

The conclusion is that we can construct the line joining two ideal points and the line joining one ideal point and one finite point by compass and finite straightedge.

Exercises

1. Let ℓ and m be ultraparallel lines. From arbitrary points A and A' on ℓ, drop perpendiculars to m, with feet at B and B', as shown in Fig. 11.9a. In case $|AB| = |A'B'|$, the common perpendicular to ℓ and m can be constructed as the perpendicular bisector of the segment AA' or the segment BB'. (See Exercise 3 in Section 2.8.)

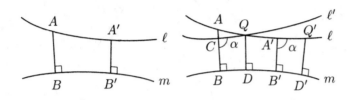

Fig. 11.9a   **Fig. 11.9b**

Otherwise, assume that $|A'B'|$ is less than $|AB|$. Construct the point C on AB such that $|CB| = |A'B'|$. Construct line ℓ' through C at the angle equal to the angle α between $A'B'$ and ℓ, as shown in Fig. 11.9b. Prove that ℓ' intersects ℓ at some point Q (either between A and A' or to the right of A'). Construct point Q' on ℓ to the right of A' so that $|A'Q'| = |CQ|$, drop perpendiculars to m from Q and Q', with feet at D and D', as shown. Show by the SASAA criterion (Exercise 1 in Section 2.8) that the quadrilaterals $CBDQ$ and $A'B'D'Q'$ are congruent, so that $|QD| = |Q'D'|$. Then the common perpendicular of ℓ and m is the perpendicular bisector of the segment QQ' or the segment DD'. According to Theorem 3.14 in Section 3.5, the common perpendicular is unique; hence the result of this construction is independent of the choice of the points A and A' on ℓ.

2. Let ℓ and ℓ' be directed nonasymptotic lines and let L and L' be the corresponding ideal points at infinity determined by ℓ and ℓ'. (There would be no harm if ℓ and ℓ' were asymptotic in the reversed directions; what is essential is that the ideal points L and L' be distinct.) We wish to construct the unique line m between L and L'. (m is asymptotic to ℓ in one direction on m and asymptotic to ℓ' in the other direction.) See Fig. 11.9c. Let A and A' be arbitrary points on ℓ and ℓ', respectively. By Bolyai's method mentioned above in the text, construct a ray $\vec{k}$ from A asymptotic to ℓ' and a ray $\vec{k}'$ from A' asymptotic to ℓ. Prove that the bisectors n and n' of the angles $\angle \vec{k}, \ell$ and $\angle \vec{k}', \ell'$ meet the line m at right angles; hence m can be constructed by the method of Exercise 1 as the common perpendicular bisector of n and n'. Since m is uniquely determined by the ideal points L and L', the result of this construction is independent of the choice of the points A and A'.

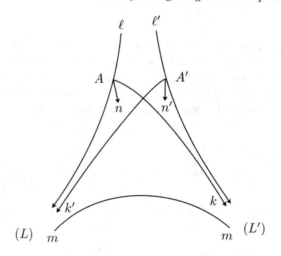

Fig. 11.9c.

3. Given three noncollinear points, show how to determine whether they lie on a circle, on a horocycle, or on an equidistant. In the first case, show how to construct the center of the circle; in the second how to determine the ideal center of the horocycle; in the third how to determine a line ℓ and a distance d such that every point of the equidistant is at a distance d from ℓ.

11.10 A Set of Axioms Omitting Completeness

From a conceptual point of view, rather than from that of a draftsman, the true significance of the theory of constructions is that the set of constructible points has all the properties of the hyperbolic plane except completeness. The real number system $\mathbb{R}$ enters in Axiom 3 in such a way that every bounded increasing set of points on a line has a limit point (on the line). We wish to eliminate limiting processes, hence we must replace $\mathbb{R}$ by some subset of the real numbers that is not necessarily complete. A similar change has to be made in connection with angles in Axiom 5.

The *incomplete hyperbolic plane* $\mathbb{I}^2$ is a subset of the set $\mathbb{H}^2$ of points in the hyperbolic plane, together with certain subsets of $\mathbb{I}^2$ called *lines*, subject to the following axioms. Henceforth, the word "line" will refer always to one of the lines in $\mathbb{I}^2$ and the word "point" will refer to one of the points in $\mathbb{I}^2$. Axioms 1 and 2 are unchanged.

Axiom 3′: There is a subset $\mathbb{S}$ of the set $\mathbb{R}$ of real numbers, closed under addition and subtraction, with the property that for each line ℓ, there is a one-to-one mapping from the set ℓ of points to the set $\mathbb{S}$ of real numbers, such that if A and B are any points on ℓ and $x(A)$ and $x(B)$ are the corresponding numbers, then $|AB| = |x(A) - x(B)|$.

The set $\mathbb{S}$ is not specified at this point, but is indirectly determined by the other axioms. It will in fact turn out to be the set of logarithms of positive numbers in quadratic-surd fields; that is, a countable set.

Axiom 4 is unchanged, but it acquires a new importance, because it tells that two lines intersect if they "ought to;" namely, if P and Q are on opposite sides of ℓ, then the segment PQ intersects ℓ. In the strictly hyperbolic plane $\mathbb{H}^2$, the existence of the intersection was a consequence of the completeness of the sets of points on the lines.

To ensure the existence of points obtained by intersecting lines and circles, we need the following.

Axiom 4a: If P and Q are on opposite sides of a line ℓ, there is a point R on ℓ such that $|PR| = |PQ|$.

R is the intersection of the line with the circle of radius $|PQ|$ centered at P. That there are just two such points R (i.e., two points of intersection of the line and the circle) then follows from the properties of the set $\mathbb{S}$.

Axiom 5 is unchanged, except that the set of all real numbers in $(0, \pi)$ is replaced by a set $\mathbb{S}'$ of real numbers in $(0, \pi)$. Again, the set $\mathbb{S}'$ is not specified here, but it will turn out, since the angle of parallelism $\Pi(y)$ can be constructed when y is given, and y can be constructed when $\Pi(y)$ is given, that the set $\mathbb{S}'$ consists $(\mathrm{mod}\,\pi)$ of the arcsines of numbers between -1 and $+1$ in quadratic-surd fields.

Axioms 6 and 7b are unchanged. One more axiom is needed.

Axiom 8: If P is a point on a line ℓ, there is a line m perpendicular to ℓ at P.

The classical construction of such a perpendicular involved the intersection of two circles. So far, we have said nothing about the existence of the point of intersection of two circles, hence the need for the axiom. From this axiom, it follows that every segment has a midpoint, and every angle has a midray.

With these axioms, it now follows that the set of constructible points in $\mathbb{H}^2$ is a minimal incomplete hyperbolic plane $\mathbb{I}^2$. All points of $\mathbb{I}^2$ can be constructed.

Since the minimal incomplete hyperbolic plane $\mathbb{I}^2$ is countable, it illustrates a theorem of Skolem and Löwenheim, which says that if a system of axioms, say for a geometry, complies with certain rules of discourse associated with first-order logic, then there is necessarily a countable model. To describe the rules of discourse in detail would take us too far afield into mathematical logic, but we may point out the following: The rules forbid one to speak about arbitrary subsets of a given set, and a completeness axiom, for example, for the real number system $\mathbb{R}$, would do just that: it would say in fact that if an arbitrary subset of $\mathbb{R}$ is bounded, then it has a least upper bound.

Exercises

1. Show how to determine the midpoint of a segment, using Axiom 8.
2. Show how to determine a ray that bisects an angle.
3. Suppose that the straightedge and compass are augmented by an instrument that can trisect a segment. What sort of more general fields do the quadratic-surd fields have to be replaced by in the construction theory?

Index

Universitext *(continued)*

Ma/Roeckner: Introduction to the Theory of (Non-Symmetric) Dirichlet Forms
MacLane/Moerdijk: Sheaves in Geometry and Logic
Marcus: Number Fields
McCarthy: Introduction to Arithmetical Functions
Meyer: Essential Mathematics for Applied Fields
Meyer-Nieberg: Banach Lattices
Mines/Richman/Ruitenburg: A Course in Constructive Algebra
Moise: Introductory Problem Course in Analysis and Topology
Montesinos: Classical Tessellations and Three Manifolds
Morris: Introduction to Game Theory
Nikulin/Shafarevich: Geometries and Groups
Øksendal: Stochastic Differential Equations
Porter/Woods: Extensions and Absolutes of Hausdorff Spaces
Ramsay/Richtmyer: Introduction to Hyperbolic Geometry
Rees: Notes on Geometry
Reisel: Elementary Theory of Metric Spaces
Rey: Introduction to Robust and Quasi-Robust Statistical Methods
Rickart: Natural Function Algebras
Rotman: Galois Theory
Rybakowski: The Homotopy Index and Partial Differential Equations
Sagan: Space-Filling Curves
Samelson: Notes on Lie Algebras
Schiff: Normal Families of Analytic and Meromorphic Functions
Shapiro: Composition Operators and Classical Function Theory
Smith: Power Series From a Computational Point of View
Smoryński: Logical Number Theory I: An Introduction
Smoryński: Self-Reference and Modal Logic
Stanišić: The Mathematical Theory of Turbulence
Stichtenoth: Algebraic Function Fields and Codes
Stillwell: Geometry of Surfaces
Stroock: An Introduction to the Theory of Large Deviations
Sunder: An Invitation to von Neumann Algebras
Tondeur: Foliations on Riemannian Manifolds
Verhulst: Nonlinear Differential Equations and Dynamical Systems
Zaanen: Continuity, Integration and Fourier Theory